CAMBRIDGE LIBRARY COLLECTION
Books of enduring scholarly value

Technology

The focus of this series is engineering, broadly construed. It covers technological innovation from a range of periods and cultures, but centres on the technological achievements of the industrial era in the West, particularly in the nineteenth century, as understood by their contemporaries. Infrastructure is one major focus, covering the building of railways and canals, bridges and tunnels, land drainage, the laying of submarine cables, and the construction of docks and lighthouses. Other key topics include developments in industrial and manufacturing fields such as mining technology, the production of iron and steel, the use of steam power, and chemical processes such as photography and textile dyes.

Life and Labours of Thomas Brassey

An important figure in British business history, the civil engineering contractor Thomas Brassey (1805–70) stood at the forefront of railway construction across the globe in the nineteenth century. He was also instrumental in the development of the Victoria Dock and part of London's sewer system. Originally published in 1872 and reissued here in its 1888 seventh edition, this first biography of Brassey was written by his personal friend, the public servant and author Sir Arthur Helps (1813–75). It describes Brassey's many remarkable achievements as a prolific contractor working in Europe, Asia, Australia and the Americas. A brilliant businessman, representing the best of British skill, leadership and organisation, Brassey employed tens of thousands of men around the world at the peak of his career. Having collaborated with prominent engineers such as Joseph Locke and Robert Stephenson, he secured for himself a long-lasting reputation for integrity and dedication.

Cambridge University Press has long been a pioneer in the reissuing of out-of-print titles from its own backlist, producing digital reprints of books that are still sought after by scholars and students but could not be reprinted economically using traditional technology. The Cambridge Library Collection extends this activity to a wider range of books which are still of importance to researchers and professionals, either for the source material they contain, or as landmarks in the history of their academic discipline.

Drawing from the world-renowned collections in the Cambridge University Library and other partner libraries, and guided by the advice of experts in each subject area, Cambridge University Press is using state-of-the-art scanning machines in its own Printing House to capture the content of each book selected for inclusion. The files are processed to give a consistently clear, crisp image, and the books finished to the high quality standard for which the Press is recognised around the world. The latest print-on-demand technology ensures that the books will remain available indefinitely, and that orders for single or multiple copies can quickly be supplied.

The Cambridge Library Collection brings back to life books of enduring scholarly value (including out-of-copyright works originally issued by other publishers) across a wide range of disciplines in the humanities and social sciences and in science and technology.

Life and Labours of Thomas Brassey

1805–1870

ARTHUR HELPS

CAMBRIDGE
UNIVERSITY PRESS

University Printing House, Cambridge, CB2 8BS, United Kingdom

Published in the United States of America by Cambridge University Press, New York

Cambridge University Press is part of the University of Cambridge.
It furthers the University's mission by disseminating knowledge in the pursuit of
education, learning and research at the highest international levels of excellence.

www.cambridge.org
Information on this title: www.cambridge.org/9781108067812

© in this compilation Cambridge University Press 2014

This edition first published 1888
This digitally printed version 2014

ISBN 978-1-108-06781-2 Paperback

This book reproduces the text of the original edition. The content and language reflect
the beliefs, practices and terminology of their time, and have not been updated.

Cambridge University Press wishes to make clear that the book, unless originally published
by Cambridge, is not being republished by, in association or collaboration with, or
with the endorsement or approval of, the original publisher or its successors in title.

BOHN'S SELECT LIBRARY

LIFE OF THOMAS BRASSEY

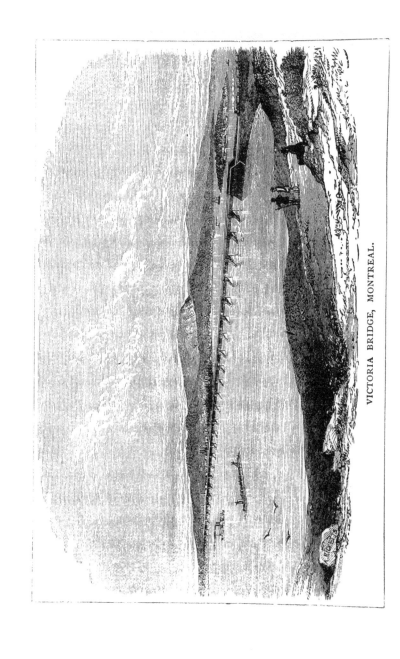

VICTORIA BRIDGE, MONTREAL.

LIFE AND LABOURS

OF

THOMAS BRASSEY

1805—1870

BY THE LATE
SIR ARTHUR HELPS, K.C.B.

SEVENTH EDITION

LONDON: GEORGE BELL AND SONS, YORK STREET
COVENT GARDEN
1888

CHISWICK PRESS: C. WHITTINGHAM AND CO., TOOKS COURT,
CHANCERY LANE.

DEDICATION.

TO THE QUEEN.

MADAM,
    I am very grateful for the permission given me to dedicate this work to Your Majesty.

I desired so to dedicate it, because I do not know of anyone who has a deeper sympathy with the labouring classes than Your Majesty, or anyone who takes a more heartfelt interest in everything that concerns their habits, their education, and their general welfare. Moreover, this sympathy and this interest are not confined to those classes in Your Majesty's Dominions only, but are extended to them wherever they are to be found.

I think also, that it cannot but be very gratifying to Your Majesty to have full evidence that, in a special kind of labour of a very important character, namely, the construction of railways, Your own subjects have hitherto borne the palm, and have introduced their excellent modes of working into various Foreign countries.

Your Majesty will find that the late MR. BRASSEY was an employer of labour after Your Majesty's own heart: always solicitous for the well-being of those who served under him; never keeping aloof from them, but using the powerful position of a master in such a manner as to win their affections, and to diminish the distance which is often far too great between the employer and the employed.

I venture, therefore, to think that the volume will be interesting to Your Majesty on its own account; and that You will be disposed to view with favour the merits, if any, and to deal gently with the faults, of a work written by one who, with all respect, is ever Your Majesty's
                Faithful and devoted
                    Subject and Servant,
                        ARTHUR HELPS.

LONDON: *June,* 1872.

PREFACE.

I FEEL it to be right to acknowledge in detail the great assistance I have received in writing this work. It would be difficult for me to name all the persons from whom I have derived this assistance; but I must mention some of those who have been my principal coadjutors; namely, Mr. Thomas Brassey, Mr. Ballard, Mr. Bidder, Dr. Burnett, Mr. Day, Mr. Dent, Mr. Edwards, Mr. Evans, Mr. Fowler, Mr. Netlam Giles, Mr. Hancox, Mr. Henry Harrison, Mr. Hawkshaw, Mr. Henfrey, Mr. Hodges, Mr. Holme, Mr. Charles Jones, Mr. Longridge, Mr. Louth, Mr. Mackay, Lieut.-Col. Charles Manby, Mr. Milroy, Mr. Frederic Murton, Mr. Ogilvie, Sir Morton Peto, Bart., Mr. Ray, Mr. Rhodes, Mr. Ricketts, Rev. H. Robinson, Mr. Rowan, Mr. Strapp, Mr. Tapp, Mr. Trubshaw, Mr. Wagstaff, Mr. Wilcox, and Mr. Woolcott.

The object of this work is not merely to narrate the life and labours of Mr. BRASSEY; but it aims, also, to show forth the labours of others, which that life elicited. The life of many an eminent man, especially if his eminence has consisted in doing one kind of work very well, does not admit of much interest in the narrative itself, and might be very briefly told. It is a melancholy fact, but fact it is, that great conquerors are mainly the persons whose lives are most interesting, such as Alexander the Great, Julius Cæsar, Cortes, and Napoleon. On the contrary, the lives that have been most serviceable to mankind, as well as the histories of the most peaceful and therefore happiest periods of the lives of nations, give little scope for exciting narrative. The consequences, however, of the actions of these benefactors of mankind are often of the highest interest; and, in this particular case of Mr. Brassey, those consequences were visible in his own lifetime, and may therefore naturally be incorporated with any account of his life.

Mr. Brassey must ever be considered as one of the chief pioneers in the great series of industrial enterprises by which the modern world has, we may almost say, been transformed. The interest in his life greatly depends upon the fact, that his career and the establishment of railways commenced almost simultaneously. He certainly was the first person who went out as a contractor into foreign lands, and who first made the British modes of working known in many parts of the globe. It was fortunate for our reputation with the foreigner, that British skill, workmanship, and power of organization, as manifested in railway construction, were made known in foreign countries by one who was a type of the men of his calling, and who possessed in perfection some of the most sterling qualities of the British people.

When I speak thus of Mr. Brassey, I do not mean for a moment to ignore the services of the engineers under whom he acted, or of the partners with whom he acted in these great undertakings. From Mr. Mackenzie, Mr. Brassey's first partner in a foreign railway contract, from many of his succeeding partners, from the able and distinguished men in their several callings who were connected with him in railway enterprises, Mr. Brassey received the greatest assistance. The business of a contractor is not by any means of an isolated character, and, in the course of Mr. Brassey's life, as will be seen by reference to the Table of Contracts in Chapter XII., he had at least twenty-seven partners. To narrate adequately the work that these gentlemen did in Great Britain and in Foreign countries would require many biographies to be written. I feel justified, however, in claiming Mr. Brassey as a representative man from whose career the great exertions and the skill manifested in railway enterprise by British engineers, contractors, agents, and workmen of all kinds, may be adequately appreciated.

Not the least valuable and interesting part of the book, if I may presume to speak of any part as valuable or interesting, is that which gives, incidentally, an account of national characteristics. Lavater said that you could not thoroughly understand a man until you had divided an inheritance with him. Without going so far, I would venture to say that you cannot thoroughly understand a man's nature until you have done business with him, for it is in the transaction of business that all the qualities of a man come forth and are developed. Mr. Brassey himself, and his agents, all of them very shrewd and capable persons, had to deal with men of every class in the countries where they were constructing railways. It was a necessary part of their business to understand the characters of the foreigners they dealt with; and the remarks of these agents show that they did not fail to accomplish that primary portion of their labours. Frenchmen, Belgians, Germans, Italians, Russians, Spaniards, and Danes came under the close observation of Mr. Brassey and his agents; and we are told how the men of these various nationalities acquitted themselves in their respective employments. Sometimes we find that our preconceived notions of the characteristics of certain peoples are confirmed: sometimes we find that these preconceived notions require modification. But whatever we learn upon this subject from persons who had such remarkable opportunities of observation, is pure gain, and tends to remove our notions from the region of prejudice to that of fact.

CONTENTS.

		PAGE
INTRODUCTORY CHAPTER	. . . . .	1
CHAP. I. Mr. Brassey as a Business Man	. . . .	5
II. Mr. Brassey's Early Career	. . . .	11
III. Contract Work	. . . .	21
IV. Commencement of Foreign Work	. . .	28
V. Labourers of Different Nations	. . . .	40
VI. Mr. Brassey's Skill in Calculations	. . .	52
VII. Work becomes more Extensive	. . . .	57
VIII. Great Northern Railway	. . . . .	63
IX. Financial Management	. . . . .	68
X. Financial Difficulties	. . . . . .	75
XI. Mr. Brassey's Wealth	. . . . . .	80
XII. Railway and other Contracts	. . . .	83
XIII. Italian Railways	. . . . . .	93
XIV. Grand Trunk Railway of Canada	. . .	101
XV. Victoria Bridge	. . . . . .	110
XVI. Crimean and other Railways	. . . .	118
XVII. Works in Australia	. . . . . .	126
XVIII. Argentine Railway	. . . . . .	132
XIX. Moldavian Railways	. . . . .	140
XX. Indian Railways	. . . .	146
XXI. Recollections of his Son	. . . . .	150
XXII. Close of Life	. . . . . .	161
XXIII. Railways and Government Control	. . .	174
APPENDIX A. Mr. Tapp's Notes on Mr. Brassey's Tours	.	181
APPENDIX B. Letters	. . . . . .	185
APPENDIX C. Wages	. . . . . .	197
INDEX	. . . .	203

LIST OF ILLUSTRATIONS.

VICTORIA BRIDGE (by *A. Ricketts*)	*Frontispiece*	
BULKELEY·HALL (by *A. Ricketts*, from a Photograph)	*page*	13
MAP SHOWING SOME OF THE CONTINENTAL CONTRACTS (by *A. Ricketts*)	,,	29
MAP OF ENGLISH LINES (by *G. Edwards* and *A. Ricketts*)	,,	59
MAP OF GRAND TRUNK RAILWAY (by *A. Ricketts*)	,,	103
MAP OF ARGENTINE RAILWAY (by *A. Ricketts*)	,,	134

LIFE AND LABOURS OF THE LATE THOMAS BRASSEY.

INTRODUCTORY CHAPTER.

THE first endeavour for a writer should be to put himself in good relation with his readers. If he fail in doing this, he fails in a most important point. They may agree with him, or differ from him, as regards his conclusions; but they will almost always have gained some profit from his work, if he makes them sympathize with him, and understand his meaning and purpose. Both reader and writer have but one and the same object: namely, to get at the truth in regard to a person, or matter, about whom, or which, it is worth while to know the truth.

In writing the life of Mr. Brassey, I have undertaken a task for which I have no special qualifications; but the aid I have received from his family and from his many friends who do possess these special qualifications, has greatly lightened my labour, and will, I hope, make it effectual in bringing before the world the character and conduct of a very notable person.

Before commencing this biography in the ordinary way, by giving the birth, parentage, and education of Mr. Brassey, I shall begin by giving an account of my first acquaintance with him—an acquaintance which afterwards ripened into a sincere friendship, causing me to accept with pleasure the task of writing his life, when requested by his sons to do so. They would, no doubt, have done the work better and more amply; but then, what a son says of his father is always a little "suspect." Notwithstanding the familiarity of converse which has grown up of late years between sons and fathers, sons are apt to be not the less proud, and perhaps even more fond, of their fathers than ever; and on that account not the less unfit to write their fathers' lives.

I am confirmed in this view, by the life of an eminent man, written by his sons, which, though very well and certainly very dutifully written, failed to give the reader an adequate notion of those peculiarities in the hero of the tale, which are so valuable in making us really acquainted with him. This knowledge the reader did not, I think, attain until he came to a letter at the end of the book, written by Sydney Smith. It was a letter which thoroughly succeeded in bringing the man before you by means of such passages as the following: "Curran, the Master of the Rolls, said to Mr. Grattan, 'You would be the greatest man of your age, Grattan, if you would buy a few yards of red tape, and tie up your bills and papers.' This was the fault, or the misfortune, of your excellent father; *he never knew the use of red tape, and was utterly unfit for the common business of life.* That a guinea represented a quantity of shillings, and that it would barter for a quantity of cloth, he was well aware; but the accurate number of the baser coin, or the just measurement of the manufactured article to which he was entitled for his gold, he could never learn, and it was impossible to teach him."[1] Now these are the kind of things which sons are too fond and too respectful to say of their fathers; and therefore I do not think that sons can ever make good biographers.

I now proceed to give an account of my first introduction to Mr. Brassey. I had to receive a visit from him on some official business of much importance and considerable difficulty. When one has heard a great deal of a man, but has not seen him, one cannot help forming some notion as to what manner of man he is.

When Mr. Brassey's name was announced, I could not help supposing that I should see a hard, stern, forcible, soldierly sort of person, accustomed to sway armies of

---

[1] "Life of Mackintosh," vol. ii. p. 500.

working men in an imperious fashion. Now this was very foolish of me; for I had, before, seen many great "captains of industry," and had almost uniformly found them to be men of suave manners and courteous bearing.

Notwithstanding this experience, I was prejudiced and misled by the word "contractor," and expected to find in Mr. Brassey a very different person from the one I did see. There entered an elderly gentleman of very dignified appearance, and of singularly graceful manners, suggesting at once the idea of what is called a "gentleman of the old school."

He stated his case. No: I express myself wrongly; he did not state his case; he *understated* it; and there are few things more attractive in a man than that he should be inclined to understate rather than to overstate his own case. He was also very brief; not going over any part of the ground a second time, as is the habit of ninety-nine persons out of every hundred. After he had gone away, I thought to myself (for I knew the matter pretty well, in respect of which he had a grievance) that, had it been my case, I should not have been able to restrain myself so completely and to speak with so little attention to self-interest as he had done.

On thinking whom he resembled of the persons I had ever seen, I found that he reminded me most of the late Lord Herbert of Lea, a man who, even in a short and transient interview, never failed to impress you with a sense of his goodness and benevolence, and of his being one of the most perfect gentlemen you had ever seen.

This was my first interview with Mr. Brassey. The impression it produced upon me was that of respect and regard for him, which continued to increase as we became better acquainted.

I have also to add, that the life of Mr. Brassey has especial interest for the writer of it, as affording an example of skilful organization, as well as of the fulfilment of other functions, and the performance of other duties, which, though they especially concern Imperial Government, may be thoroughly exemplified in the conduct of private enterprise, when it assumes considerable magnitude and diversity.

This work has been written in a very peculiar manner. Most of the persons who knew Mr. Brassey well, who had acted with him, or served under him, have kindly consented to be examined as witnesses, and to have their evidence taken down by a shorthand writer. Mr. Thomas Brassey has been the examiner. From his general knowledge of his father's affairs, no one could have fulfilled this office so well; and I gratefully acknowledge the immense assistance that I have derived from the mode in which he has conducted these examinations.

It may easily be conjectured that the amount of material thus collected has been very great indeed, and has been of the most interesting character. I have often regretted that want of space prevents me from giving to my readers as much as I should otherwise like to give of this valuable information.

CHAPTER I.

A BRIEF OUTLINE OF MR. BRASSEY'S CHARACTER AS A MAN OF BUSINESS.

Trustfulness in his agents.—Liberality and equanimity.—Powers of organization.—Delicacy in blaming.—Courtesy.—Presence of mind.—Hatred of contention.—Anxiety to have work well done.—Gangers taken into council.—Ruling passion.

IN a biography, it is a difficult matter to determine where one should introduce a description of the character of the person written about. I have come to the conclusion that the best plan is to give very early in the book a brief outline; then, as the occasions arise, to point out, in the narrative, illustrations of the character; and, finally, to take an opportunity of restating and enlarging the description.

The most striking point in Mr. Brassey's character, and that which I shall mention first, was his trustfulness. This virtue was carried to a great extent in him,—to an extent that may appear almost extreme. He chose his agents with great care, and with consummate judgment. After he had chosen them, he placed implicit trust in them. Then, though perfectly capable of exercising the most minute supervision and criticism of details, he never judged by details, but looked to results; not vexing or wearying those who served under him by minute and tiresome criticism.

He was exceedingly liberal in the conduct of his business, as will be seen from many instances in the following pages; and probably there never was a man who made so much money, caring so little for the money itself.

He was a man of a singularly calm and equable temperament. It was very rarely, indeed, that either success or failure—and even great failure was not a thing unknown to him—discomposed his complete serenity of mind. I do not mention this by way of praise. As regards this matter, there are two orders of men. There is the man whose anxieties never leave him, who cannot throw off his robe of office and say "Lie there, Lord Treasurer, or Lord Chancellor." And there is the man who, having done his best, is satisfied with that best, and can dismiss anxiety as to the result. This is a great felicity of temperament. Those men who do possess it are often liable to be much misconstrued. The world is apt to think that the man who can throw off the burden of care, is, on that account, less caretaking than the man who is harrowed by perpetual anxiety, and who cannot conceal the constant pressure of that anxiety. Mr. Brassey did not take less care than these anxious men are wont to do; but, having given his best efforts to ensure success, was content to await the result, and to abide by it with perfect equanimity.

It is not requisite to do more here than to allude to the powers of perception, of calculation, and of organization, which Mr. Brassey possessed. These powers will inevitably reveal themselves in the course of the narrative, and may, indeed, almost be taken for granted as belonging to one who successfully carried out great undertakings in which these powers were absolutely indispensable.

There was not anything more noticeable in Mr. Brassey's conduct of business than his mode of blaming where blame was requisite. It was of the very lightest and gentlest kind; but not on that account less forcible or less instructive. To speak metaphorically, his little finger laid gently upon an error was more severely felt than the heavy hand so often put down by a coarse man when he blames his agents or his inferiors. Reluctant blame is the blame that goes to the hearts and consciences of men; and the greatest merit of it is, that while it condemns, it does not discourage.

So thoroughly beloved, and so thoroughly appreciated was Mr. Brassey by all the people who served under him, that his coming amongst them was looked forward to as a most joyful and festive event. When, for instance, he had any great work on hand in a foreign country, the thousands of people employed by him, from the highest to the lowest, longed to see him amongst them. This could not have been the case had he not been utterly devoid of captiousness, and one of those generous employers of labour who recognize to the full all that is well done by those who work under them.

Indeed, in this respect, he reversed the relative positions of employer and employed. When any disaster occurred on the works, it was *he* who comforted and excused his agents, instead of receiving comfort or excuses from them.

It was a necessity of Mr. Brassey's career that he should live much with his dependents. Now, it may often be observed that the man who has undoubted authority over his fellow-men in one respect, is apt to endeavour to extend that authority to matters in which he has not any right whatever to interfere with those inferiors, or, otherwise than indirectly, to attempt to influence their opinions. The uniform testimony of those who, in any capacity, worked under Mr. Brassey, is, that he never sought to interfere with them, or their opinions, " out of school " as we may say. He was one of the least arrogant of men in his general converse with mankind, giving a respectful consideration to whatever anyone had to say to him. Even if people talked folly to him, his comment upon it was of the mildest kind. Once, indeed, when a man was talking largely, with very little substance or understanding in his talk, Mr. Brassey was heard to remark, " I think the peas are over-growing the sticks." But this was a rare instance of censure—so rare that it greatly attracted the attention of the hearers.

Mr. Brassey was gifted with much presence of mind. The first Napoleon used to say of himself, that few men were his equals in what he was wont to call " two o'clock of the morning " courage, which is in fact presence of mind on the announcement of unexpected danger and difficulty. Mr. Brassey was fortunate enough to possess this " two o'clock of the morning " courage in a high degree. If called up suddenly in the middle of the night upon some urgent peril or difficulty, he met the alarm with perfect coolness; sat down to consider and calculate what was the best mode of obviating the danger (danger seemed to stimulate his faculties, and not to overpower them); and, before the break of day, when he had to proceed to the scene of action, was ready with his plan. It may be easily imagined what confidence this presence of mind on the part of their employer, infused into his principal agents, and all those who were employed under him.

Mr. Brassey had a perfect hatred of contention. This quality of mind was, second only to his trustfulness, the main element of his success. It was soon discovered by anyone who had dealings with him that, should any matter of controversy arise, he would not only refuse to take any questionable advantage over the other side, but would rather even submit to be taken advantage of.[1] Now, there is not a more fruitful virtue in the world than this kind of generosity. It is nearly sure to elicit a kindred response. In most instances where overreaching is begun or continued, it derives its strength from contentiousness.

In the execution of any great undertaking Mr. Brassey's anxiety was that the work should be done quickly, and be done well. The minor questions as to who should bear the expense of minor matters, unprovided for by specific contract, he left to be settled afterwards; whereas, many men, perhaps I may say most men, would have insisted, beforehand, upon the question being settled as to who should bear the outlay. Mr. Brassey's name is a name not known in the Law Courts. He said to Mr. Giles one day: "I never had but one regular law-suit. It was in Spain about the Mataro Line, and that was against my will; but I was obliged to submit to it, as I had a partner. We got nothing by it; and I will never have another if I can help it, for I believe in nineteen cases out of twenty you either gain nothing at all, or what you do gain does not compensate you for the worry and anxiety the law-suit occasions you." If a dispute arose between his agents and the engineers of the company, for whom he was working, as to the best mode of proceeding with the work, he had an admirable way of settling the dispute. He would appear, perhaps unexpectedly, amongst the contending parties; would not back up his own agents, or enter into vexatious contention with the engineers of the company; but would, in the presence

---

[1] See Letter No. 1, in Appendix.

of them all, take the "gangers" into council, and ask them what was their opinion on the matter.

It was generally found that the gangers had a very clear opinion, and a very judicious one, as to how the work should proceed: and, at any rate, the contending parties felt that the opinion of those men, with whom the manual execution of the work rested, was an opinion which it was very desirable to defer to and to conciliate. This mode of reference and undefined arbitration was eminently characteristic of this great employer of labour. It did not vex or humiliate anybody; and it brought the matter to a definite conclusion.

Our immediate forefathers, in estimating the character of any man, were always anxious to point out what was his ruling passion. This may be seen in the poets of a former age. They could not conceive the idea of a man unswayed by a ruling passion, which indeed they would invent for him, if he were not blessed, or cursed, with such a motive for endeavour. I must confess that I think the idea is not altogether a bad one, and that most men have a ruling passion—strong in life, as in death. Now, in writing this memoir, I have endeavoured to find out what was Mr. Brassey's ruling passion; what was the work that he, Mr. Brassey, supposed that he was sent into the world to further and to establish. He had none of the ordinary ambitions. Rank, title, social position had no attraction for him. He had no other objects than those connected with his business. His great ambition—his ruling passion, if I may so express it—was to win a high reputation for skill, integrity, and success in the difficult vocation of a contractor for public works; to give large employment to his fellow-countrymen; and by means of British labour and British skill to knit together foreign countries, and to promote civilization, according to his view of it, throughout the world.

Mr. Brassey was, in brief, a singularly trustful, generous, large-hearted, dexterous, ruling kind of personage; blessed with a felicitous temperament for bearing the responsibility of great affairs.

By giving at once this view which I have formed of Mr. Brassey's character, I hope I may have sufficiently in-

terested the reader to induce him to accompany me on my
journey through the details, sometimes of a dry and technical character, which serve to illustrate the nature of a
man who undoubtedly proved himself to be one of the
foremost leaders of industry in the present age.

CHAPTER II.

MR. BRASSEY'S EARLY CAREER.

(A.D. 1805-1837.)

Birth and parentage.—Goes to school, and is articled.—The Holyhead Road.—Becomes Mr. Lawton's partner.—Birkenhead in 1818.—Mr. Price's agent.—Mr. Stephenson.—Stourton Quarries.—The Sankey Viaduct.—First tender for railway works.—The Dutton Viaduct.—First contract.—Difficulties of early railway-making.—Meets Mr. Locke.—London and Southampton Railway.—Marriage.—Mr. Harrison of Birkenhead.—Mrs. Brassey.—Early objections to railways.

MR. BRASSEY was born November 7, 1805, at Buerton, in the parish of Aldford, in Cheshire. He was the son of John and Elizabeth Brassey of that parish.

His family was an ancient one, his ancestors having come over with William the Conqueror. For nearly six centuries they resided at Bulkeley, near Malpas, in Cheshire, where they possessed a small landed property of three or four hundred acres, which is still in the family. Mr. Brassey was much attached to this ancestral property, and when the old house became almost uninhabitable from the effects of time, he rebuilt upon its site a handsome house, with model farm buildings on a large scale.

Like most other ancient families, the Brasseys were concerned, in one way or another, in the civil war of the Roses; but, whatever losses they may have sustained at that period, so disastrous to many ancient English families, they were fortunate enough to retain a large part of their property.

The time when they moved to Buerton is uncertain; but they must have resided there for more than two centuries, as is proved by certain documents which are dated in the year 1663. Mr. Brassey's father, in addition to property which he possessed in Cheshire, had land of his own at Buerton, and rented from the Marquis of Westminster a large farm adjacent to it. The rent of this farm was £850 a year.

I am particular in noting these facts about the history of Mr. Brassey's family, because it resembles that of many of those families from which our most distinguished men have sprung—an origin which I conceive is very favourable for a man who is destined to do great things in this world. There is a certain amount of culture and of knowledge in such a family; while at the same time it has run no risk of being enervated by luxury, or of having, if I may venture to use the expression, thought itself out. We cannot be blind to the fact that there are amongst us but few descendants of our most eminent men. It certainly seems as though a family, after long ages, like some slowly developing plant, produces its best flower, and then dies off. And when we see distinguished families still producing remarkable men, I believe that if we could investigate the records of those families, we should find that there had been a frequent accession of new blood,—of minds unwearied by mental labour, of bodies not exhausted or rendered unfruitful by luxury.

Mr. Brassey, at twelve years of age, went to a school at Chester, of which the late Mr. Harlings was master. At sixteen years of age he left school, and was articled to the late Mr. Lawton, a Land-surveyor and Agent. Mr. Lawton was at that time, and for many years had been, the agent of the late Mr. Francis Richard Price, of Bryn-y-pys, Overton, Flintshire.

Those of my readers who are no longer young, may remember that in the days of their youth there was a road made, which it was delightful to travel on, and of which all England was very proud. It was called the Holyhead Road. It commenced at Shrewsbury and terminated at Holyhead; and this was the first great work upon which the young Brassey was employed. The celebrated Telford was the engineer of this road, and under him, as a surveyor, a Mr. Penson, of Oswestry, was employed to make the surveys for the road. Mr. Brassey was permitted by his master to assist Mr. Penson in making these surveys.

BULKELEY OLD HALL.

Throughout his life we uniformly find that Mr. Brassey was a favourite with those with whom, or under whom, he acted. His master, Mr. Lawton, appreciating his value, became much attached to him, and ultimately proposed to take him into the business as a partner.

There was an additional reason, at that time, for making such a proposal. Mr. Lawton had the shrewdness to perceive that Birkenhead would become a very great place, and that it would give much occupation for men of his calling. Accordingly, he resolved to establish a business there, and to place at the head of it his young friend Brassey.

Mr. Brassey accepted the proposal of partnership; and, being then, twenty-one years of age, went to reside at Birkenhead as Mr. Lawton's partner, the whole of the Birkenhead estate being the property of Mr. Price.

Birkenhead must at that time have been but a very small place, for we have evidence that in 1818 it consisted of only four houses.

At this early period Mr. Brassey showed that ingenuity and fertility of resource which was afterwards so largely developed. At this time he possessed brick-yards and lime-kilns. In loading and unloading the barges, he found much injury occurred to the bricks: he therefore devised a kind of wooden crate, which not only kept the bricks from injury, but also reduced greatly the cost of delivery and re-stacking.

On the death of Mr. Lawton, young Mr. Brassey became the sole agent and representative of Mr. Price; and, no doubt, acquired great experience in directing for him the rise and progress of that now most populous and thriving place.

He had resided there eight years, when, accidentally, he came in contact with a great man—a circumstance which gave the colour and direction to his future life. This great man was George Stephenson.

Mr. Brassey either possessed himself, or as a land-agent had the management of, a certain stone quarry at Stourton. Stone was wanted for the Sankey Viaduct on the Manchester and Liverpool Railway—the first railway, for passenger traffic, that was ever constructed.

Mr. Stephenson went with Mr. Brassey to examine the stone at this quarry, intending, if satisfied with it, to make a contract for its delivery at the Sankey Viaduct. It is evident that Mr. Stephenson must have been much pleased with his young companion in this excursion, for he immediately sought to engage him in the new enterprise of railway-making. Acting under Mr. Stephenson's advice, Mr. Brassey was induced to tender for a contract on the Grand Junction Line. This railway was to run from Newton to Birmingham; and it now forms part of the London and North-Western system.

The first tender Mr. Brassey made was for the Dutton Viaduct, near Warrington; but his estimate did not partake of the rashness of youth, for it was £5,000 higher than the estimate sent in by the late Mr. Macintosh, a well-known contractor of that day, who obtained the contract in question.

Not discouraged by this failure, Mr. Brassey next tendered for the Penkridge Viaduct, which is between Stafford and Wolverhampton; also including in his tender ten miles of railway on the same line. Mr. Brassey was enabled to tender for this contract by the liberality of his bankers at Chester—Messrs. Dixons—who, on being informed of the circumstances, at once agreed to place a considerable sum to his credit. Mr. Brassey never forgot this act of kindness on their part, and kept his principal account with these bankers throughout his lifetime. Mr. Brassey was successful in obtaining this contract. He was now twenty-nine years of age. Doubtless he had obtained much knowledge of all kinds of construction connected with his business of a land-surveyor. The construction of railways, however, was at that time altogether a novelty, not only to him, but to all persons engaged in it. The work had not yet begun to run in grooves, after which everything is comparatively so easy; but it required new modes of operation, and the creation of skilled labour of a new kind; also the management of larger bodies of men than hitherto had been brought together for public works, and a more rapid movement of these *armies* of labouring men, from place to place, than hitherto had ever been requisite. Moreover, and this is a most important point, the system of "sub-contracts" had not been devised, or, rather, had only been partially and slightly adopted,—a system which has given increased facility to all great public works.

Altogether, to take a railway contract in those days, and to work it out successfully, was no light undertaking, but one which taxed to the utmost the ability of every kind possessed by the contractor.

Mr. Brassey having obtained this contract, completed it most successfully. Mr. George Stephenson was Engineer-in-chief when Mr. Brassey took this contract. A few months, however, after the commencement of the line, Mr. Stephenson resigned his appointment; and the late Mr. Locke, who had been his pupil and assistant, was appointed to succeed him as Engineer-in-chief to the line.

On the completion of the Grand Junction Railway, Mr. Locke was employed on the London and Southampton Railway, which had been commenced under the superintendence of the late Mr. Francis Giles.

Mr. Locke asked Mr. Brassey to go with him; and Mr. Brassey contracted for, and undertook the important works on that railway between Basingstoke and Winchester, and also on other parts of that line.

It was when Mr. Brassey was thirty-one years of age, that he came up to London, in consequence of his connection with the London and Southampton Railway, and thus entered into a much larger sphere of business; in fact, commencing a career which was to lead him into great railway operations, extending over a large part of Europe, India, and the British Possessions in America.

It may be requisite here to say something of the business relations of Mr. Locke and Mr. Brassey. It has been thought by some persons that Mr. Locke showed a spirit of favouritism for Mr. Brassey; and this is so far true, that Mr. Locke was always delighted to have Mr. Brassey as a coadjutor: but those who knew anything of the qualities of that eminent engineer, Mr. Locke, must be well aware that his regard as a man of business for any other man of business would have been founded upon no prejudices, and upon no unreasonable favouritism. To put the matter very plainly, it was soon discovered that whenever Mr. Brassey had undertaken a contract on a line, the Engineer-in-chief had but little occasion for rigid supervision. Mr. Locke well knew that a bargain once concluded with Mr. Brassey would be exactly, I may say handsomely, fulfilled, and that no difficulties or contingencies would be made an excuse for delay, or an occasion for demanding any alteration in the terms of the contract. After the fall of a certain great viaduct, which disaster will have to be mentioned in these pages, it was suggested to Mr. Brassey that, on his representing the facts of the case to the Directors of the Company, some alleviation of his loss might be obtained. His reply to this suggestion was in consonance with the whole tenour of his career. "No," he said, "I have contracted to make and maintain the road, and nothing shall prevent Thomas Brassey from being as good as his word."

Throughout Mr. Brassey's career, his faithfulness, his desire to do his work efficiently, whether at a gain or a loss, together with his resolution to avoid all petty subjects of dispute, naturally made him a most welcome fellow-worker to any person placed in such an arduous position— a position requiring so much watchfulness and supervision —as that of Engineer-in-chief to a railway. It was an immense comfort to have a man to deal with, whom it was not necessary to be looking after in respect of any of the details of the work entrusted to him.

Mr. Brassey married, on December 27, 1831, Maria, second daughter of Mr. Joseph Harrison, of Birkenhead. Mr. Harrison carried on the business of a forwarding agent in Liverpool, and acted in this capacity for the great firms of Phillips and Son, Sir J. Potter, the Houldsworths, and other leading Manchester houses. In those days there were no railways, and the business of a forwarding agent was of an important and interesting character, as he was the medium of communication between the manufacturer and the shipper. Mr. Harrison was the first resident in the new town of Birkenhead. He was a man of much intelligence and foresight. Amongst his other agencies, he acted for the old Quay Canal Company, and was one of the few persons clear-sighted sufficiently to perceive that canal property would not be ruined by the new mode of transit by railway; but that, in most instances, there would be ample employment for carriers by canal as well as by railway. Indeed, he gave evidence in favour of the Liverpool and Manchester Railway Company, at the time when they were seeking to obtain their bill in Parliament.

There was much sympathy between the father-in-law and the son-in-law; and, from the time Mr. Brassey commenced business at Birkenhead, Mr. Harrison predicted his successful career. Mr. Brassey became acquainted with his future father-in-law shortly after the time when Mr. Lawton received Mr. Brassey as an articled pupil.

Mrs. Brassey has survived her husband. It is always a difficult matter to speak in praise of those who are living, and who may not like to read commendation of themselves. But, notwithstanding this necessary reserve, it is right to mention the fact that Mr. Brassey's first connection with railways was partly due to the advice which he received from his wife. He naturally hesitated to leave Birkenhead, where he had established a large and increasing business; but his wife's spirit and sound judgment convinced her that her husband would be able to find a far more important sphere, for the exercise of his great abilities, by enlisting in the small band of men who had at that time taken in hand the construction of railways. This may now seem a thoroughly self-evident proposition; but it was not so then. My readers must bear in mind the objections that were raised to railways, even in Committees of the House of Commons. "How would the carriages ever get up hill? how would they ever be able to stop, when going down hill? what would happen if a cow were to come in the way?" Such were the agitating questions asked by the opponents of the new mode of locomotion. In short, they maintained that "these new-fangled concerns, might do to convey heavy goods (as improved tramways), but as for carrying passengers, with any comfort or safety, that was a ludicrous supposition." It was a very courageous thing for any woman to hold a contrary opinion, and to hold it so firmly that she should venture to advise her husband to throw in his fortunes with the new and much depreciated class of enterprise.

Mr. Brassey followed his wife's advice, and to her therefore is due, in no slight degree, the successful career of this remarkable man.

There is the more credit due to Mrs. Brassey, as she doubtless foresaw that, in a domestic point of view, her husband's railway engagements would impose a great burden upon her, and a burden of the kind which women especially dislike. The railway contracts, in which Mr. Brassey successively engaged, compelled repeated changes of residence. In the course of thirteen years, dating from the commencement of his career as a Railway Contractor, Mr. Brassey changed his residence eleven times: namely, from Birkenhead to Stafford; from Stafford to Kingston-on-Thames; thence to Popham Lane, in Hampshire; afterwards to Winchester; from Winchester to Fareham; from Fareham to Vernon in Normandy; from Vernon to Rouen; from Rouen to Paris; afterwards back again to Rouen; then to Kingston-on-Thames, and finally to Lowndes Square, London. It may be seen by this that the life of a railway contractor has some drawbacks, such as probably may not have been thought of by my readers.

The labour and difficulty occasioned to Mrs. Brassey, by these frequent removals, all the arrangements of which were entirely left to her, may well be imagined by those who have had any similar troubles to encounter. Many of these removals, especially those from England, necessitated frequent sales of furniture, and the most cherished articles were compulsorily parted with. These, however, though serious troubles in themselves, were much added to by the social difficulties which occur in such cases, it being very difficult, if not impossible, to make friends, or at any rate to retain them, in any neighbourhood in which a family does not reside more than a year.

Moreover, Mr. Brassey's occupation, and the remote distance of many of his contracts from his place of residence, made constant absence from home inevitable; and even when at home, little could be seen by his family of the head of the house, as he was generally absent from nine or half-past nine in the morning till ten o'clock in the evening.

Of course, the education of the children at this period rested entirely with Mrs. Brassey; and, during these years of isolation, she devoted herself with the utmost affection to the care of her sons.

Mrs. Brassey could speak French fluently, which was a great assistance to her husband when they first went to France. He never had time to acquire a command of any foreign language, though, I believe, he succeeded in contriving to understand a good many of those technical terms which it was desirable for him to master.

Notwithstanding Mrs. Brassey's domestic avocations, which, as we have seen, were large and constant, she did not fail to take the greatest interest in her husband's public career; and he was wont to take counsel with her in all the weightier matters of business in which he was concerned.

CHAPTER III.

CONTRACT WORK.

(A.D. 1838.)

Mr. Hawkshaw on railways.—Need for contractors.—Sub-contracting.—Governments and contracts.—Mr. Henfrey's speech.—Mr. Brassey's dealings with the sub-contractors.—Cost of inspections.—Co-operative system.—Butty-gangs.

PREVIOUSLY to narrating the series of great works of construction, in which Mr. Brassey was engaged in foreign countries, it may be well to give a little forethought to the nature of Contract Work, and to see why such a person was wanted at this particular juncture, and for this particular kind of work.

Mr. Hawkshaw, the eminent engineer, justly observes that " with the commencement of the railway system began an age of great works, during which undertakings of far more colossal dimensions were rapidly projected, and required to be as rapidly carried into execution. The extension of the railway system called for larger docks and larger harbours, and since the construction of the Liverpool and Manchester Railway the public works that have been executed in the United Kingdom alone far exceed all that had been done before."[1]

At first sight, it might appear that there was no need for the contractor. In early days, and perhaps we may say in comparatively barbarous times, great works were doubtless executed without the intervention of the contractor. He is an inevitable product of civilization; for, inevitably, with civilization comes the division of labour.

It is obvious that one of the main advantages of doing

[1] See Letter No. 7, in Appendix.

work by contract, and through the agency of a practised contractor, is that you command the knowledge and experience of a trained body of men, fitted for the especial work. Something of this kind is to be observed in great works which were executed before the word "contractor" came into being. We know but little, comparatively speaking, of the history of the construction of the grand cathedrals which have arisen throughout the Christian world. But we are told by those who have most carefully investigated such matters—that bands of skilled workmen, chiefly Italians, were employed in these buildings, who went from place to place, and were enabled to give instruction to the local workmen. This, in some measure, accounts for the uniformity of style pervading certain periods of church architecture.

It might be said, that at first sight, it does not appear why those who have undertaken a great public work, should not execute it by means of their own officers; but almost all experience shows that this would not be a wise course. There are many reasons for this conclusion, and some of them are based upon much knowledge of human nature. If the projectors of the undertaking had to execute the work themselves, they would undoubtedly be seduced into frequent change of plan, from the very fact of their being masters of the situation. This may be seen in much smaller matters than public works. When a gentleman undertakes to build a house for himself, many are the changes he is induced to make during the building of that house, and large is the additional expense which he generally incurs.

The system of contracting for great works necessitates much forethought before commencing them, and, for the most part, a rigid adherence to the plans originally laid down.

It may serve to show the need and value of this system of contracting for great works, that it perpetually tends to extend itself. The great contractor finds it to his advantage to sub-let portions of his contract; and these are again sub-let to smaller men, by which means individual skill and exertion are gradually developed to the uttermost. Perhaps in no way could this individual skill and exertion be more amply developed, and the merits of individual men be brought more prominently forward, than by this system of contracting. Moreover, it developes the power of bearing responsibility, and tends to create masters out of men.

Thus far as regards the advantages to be derived from doing work by contract. It must, however, be remembered that there are limits within which this practice should be confined, and that there are cases to which it is altogether inapplicable.

The demands, for example, of Government are not of a nature to be wholly met by dealings with contractors. These demands are liable to be sudden, large, imperative, and indefinite. Hence Government will do wisely not to abandon their establishments. It must also be remembered, that by keeping up some of those establishments, they have, on a sudden pressure, some of the benefits of competition within their reach, and are not entirely at the mercy of what we may call the foreign element of contracting.

So, again, as regards matter into which taste and beauty of design enter, or where there is great probability of a change of plan, the system of contracting is not altogether admirable. Even as regards that instance which I took before, of a house built by a private individual for himself, there is considerable probability that the house will be better built, at any rate more suitably built, for the owner, if it is not built by contract. But no one will contend that it is not likely to be an expensively built house.

Those works, however—whether devised by governments, public bodies, or private individuals—of which the character is definite, and the time for executing them not imperatively short, are the works which it is most advisable to have executed by contract.

Now the construction of railways is work which exactly fulfils these conditions. It is very definite in character: it is not like a demand for arms, or other warlike material, which may be wanted by a Government at a very brief notice: and, in short, it is one of those products of labour, in which contract work may be most usefully employed.

Accordingly, the contractor makes his appearance on the stage as he is imperatively required.

The practice of doing works by contract is likely to increase extensively; and therefore, it will not be without advantage to observe, in the course of this memoir, how contractors have been wholly entrusted, and with the best results, with the care of some of the greatest undertakings of our time.

It may also be noted that, in carrying out works in foreign countries, great benefit has accrued both to those countries themselves, and to the country in which these projects originated, from the works being confided to contractors who carried hither and thither bands of skilful workmen; and who, indirectly, brought much profit to the mother country, while gradually they instructed the natives of other countries in skilled labour, and made them more useful citizens than they were before. There are some parts of Europe where the condition of the whole labouring population has been permanently raised by the introduction of British skill and British labour in the execution of a particular work. And this would hardly have been the case, or at any rate would not so soon have been the case, but for the presence of the British contractor and his accompanying army of British workmen; bringing new tools, new modes of working, new methods of payment; and, in short, introducing an element of vigour and prosperity which could not have been so well introduced in any other way.

In a speech made by Mr. Henfrey, on the opening of the Meerut and Umballa section of the Delhi Railway, he says:—"How greatly the working classes of this country (India) have profited by the construction of railways may be judged by the fact, that out of the seventy-five, or eighty millions sterling expended to the present time on Indian railways, nearly two-thirds, or between forty and fifty millions, must have passed, I cannot say into the pockets, but into the hands of the working classes."

When Mr. Brassey took any contract, he let out portions of the work to sub-contractors. His way of dealing with them was this: he generally furnished all the materials, and all the plant. I find him on one occasion ordering as many as 2,400 wagons from Messrs. Ransome and May. He also provided the horses. The sub-contractors contracted for the manual labour alone.

But even Mr. Brassey, with all his intelligence and all his knowledge, could not make that intelligence and that knowledge equivalent to the minute care and daily supervision which every man exercises over matters which are completely within his control. For instance, he found that in France to provide horses was an undertaking which did not pay; and afterwards he made all the sub-contractors find their own horses. At first this could not well have been done. Bringing into a strange country Englishmen, hardly any of whom could speak the language, Mr. Brassey very properly took upon himself the greater part of the responsibility, leaving the sub-contractors to find the manual labour, and to execute the work at so much per mètre. Ultimately, the sub-contractors found the horses, the manual labour, grease for the wagons, and their own blacksmiths.

Mr. Brassey's mode of dealing with the sub-contractors was of an unusual kind, and such as could not have been adopted except by a man who had great experience of all kinds of manual work, and who was also a very just man. They did not exactly contract with him, but he appointed to them their work, telling them what price he should give for it. All the evidence I have before me shows that they were content to take the work at his price, and that they never questioned his accuracy. One of his sub-contractors thus describes the process. "They did not ask him any question. He said, 'There is a piece of work for you. Will you go into that? You will have so much for it.' And then they accepted it, and went to work."

It may somewhat surprise the reader to find that all these sub-contractors were so willing at once to accept Mr. Brassey's terms; but this is easily to be accounted for by the conviction which each of them had that, if any mistake had been made, especially a mistake to their injury, there was a court of appeal which listened very readily to any grievance, and took care to remedy it. The truth is, that Mr. Brassey would always increase the price of the contract, or make it up to the sub-contractor in some other way, if the original contract had proved to be too hard a bargain for the sub-contractor. Frequently the work appointed to the sub-contractor turned out to be of a more difficult nature than had been anticipated. He however, would not desist from the work on that account, nor make any appeal in writing to his employer. He would wait until the time when Mr. Brassey should come round to visit the works. This was generally, at this period, once or twice a month. Of course Mr. Brassey had agents who represented him, providing the necessary materials, making payments, and watching the work of the sub-contractors. These agents, however, seldom felt disposed, or were not authorised, to add to the price already agreed upon between Mr. Brassey and any of the sub-contractors.[1] The sub-contractor, therefore, who had made but an indifferent bargain, awaited eagerly the coming of Mr. Brassey to the works. One of these occasions is thus described :—

> He came, and saw how matters stood, and invariably satisfied the man. If a cutting, taken to be clay, turned out after a very short time to be rock, the sub-contractor would be getting disheartened ; yet he still persevered, looking to the time when Mr. Brassey should come. He came, walking along the line as usual, with a number of followers, and on coming to the cutting he looked round, counted the number of wagons at the work, scanned the cutting, and took stock of the nature of the stuff. " This is very hard," said he to the sub-contractor. " Yes, it is a pretty deal harder than I bargained for." Mr. Brassey would linger behind, allowing the others to go on, and then commenced the following conversation. " What is your price for this cutting ? " " So much a yard, sir." " It is very evident that you are not getting it out for that price. Have you asked for any advance to be made to you for this rock ? " " Yes, sir, but I can make no sense of them." " If you say that your price is so much, it is quite clear that you do not do it for that. I am glad that you have persevered with it, but I shall not alter your price; it must remain as it is, but the rock must be measured for you twice; will that do for you ? " " Yes, very well indeed, and I am very much obliged to you, sir." " Very well; go on ; you have done well in persevering, and I shall look to you again."

The same witness states that one of these visits of inspection would often "cost Mr. Brassey a thousand pounds; and as he went along the line in these inspec-

---

[1] This statement applies only to the earlier period of Mr. Brassey's career

tions, he remembered even the navvies, and saluted them by their names."

In making sub-contracts, Mr. Brassey was very careful to apportion them according to the abilities and experience of the sub-contractor. For example, he never liked to let the brick-work and earth-work to one man. He would let the brick-work to a bricklayer, and the earth-work to a man specially acquainted with that branch. "I have often," says one of his *employés*, "heard him mention, as a principle of action—'Each one to his own speciality.'"

Before leaving the subject of Mr. Brassey's dealings with his sub-contractors, it will be desirable to see what extent of work was, as a general rule, entrusted to a sub-contractor, and what number of men he had under him. I find that the sub-contracts varied from £5,000 to £25,000; and that the number of men employed upon them would be from one to three hundred—the former number being more common than the latter. There were also, occasionally, sub-lettings made by these sub-contractors; but this was a practice of which Mr. Brassey did not approve.

It may be remarked, that, throughout his career, Mr. Brassey favoured and furthered the co-operative system; constantly giving a certain share of the profits to his agents, and thus making them partakers in the success or failure of the enterprise. He also approved of the "butty-gang" system. This word "butty-gang" requires some explanation. It means that certain work is let to a gang of about ten or thirteen men, as the case may be, and that the proceeds of the work are equally divided amongst them, something extra being allowed to the head man. This system was originated when the formation of canals first began in England. "Butty-gangs" were afterwards employed on the Paris fortifications which were constructed by French workmen.

CHAPTER IV.

COMMENCEMENT OF FOREIGN WORK.

(A.D. 1841.)

Railway development.—France and England.—Paris and Rouen Railway.—Mr. Brassey as a master.—English abroad.—Navvy language.—Interpreters.—Wages in Belgium.—Rouen and Havre Railway.—The Barentin Viaduct.

GREAT Britain has been a fortunate country in many respects, but in hardly any more fortunate than in the time at which railway communication was introduced into the country. It was not until after great attention had been paid for many years to the subject of locomotion, and after roads and canals had been brought to a state of comparative perfection throughout England, if not throughout Great Britain, that railways began to be thought of. This was an immense advantage for this country; and it was one which was possessed almost exclusively by Great Britain. In warlike preparations it is discouraging to notice that when some kind of work has been brought to perfection, such as a stately man-of-war, or a well-constructed and well-armed battery,—that is, "well constructed and well armed" according to the latest knowledge of scientific men of that particular time,—some, comparatively speaking, small advance in science or improvement in manufacture, renders the stately ship or the well-appointed battery useless, and, as we almost say, ridiculous. It is not so, however, with what may be called the minor modes of locomotion. That a country should be traversed by these in every direction is of the greatest advantage as forming the means of easy approach to the great lines of railway.

France, as indeed almost every other continental

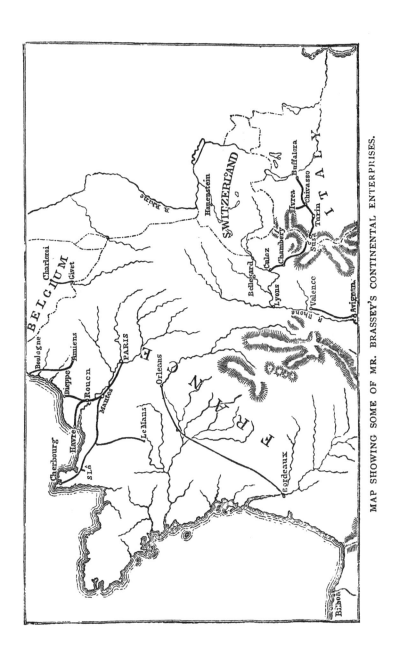

MAP SHOWING SOME OF MR. BRASSEY'S CONTINENTAL ENTERPRISES.

country, was not in the same happy position in this respect as England, at the time when railways were commenced. It was not, however, to be expected that continental countries would wait to develope a good system of roads and canals before they gave their minds to endeavouring to bring amongst them this new and marvellous mode of transit by railway, which had proved so successful and was in full activity of construction on this side of the water.

Accordingly, about the year 1830, the French began to desire that railways should be introduced into France. As a purely commercial speculation, however, there was but little hope of the railway system being adopted in that country; and consequently, the government was induced to choose, in many cases, the plan of giving guarantees to those who were willing to become shareholders in any railway undertaking.

It was natural that one of the first railways that would be thought of, was one which should connect Paris with London. Hence arose the Paris and Rouen Railway. The promoters of that line put themselves in communication with the directors of the London and Southampton Railway, and an endeavour was made to facilitate matters by an amalgamation of interests.

The proposition which the French Board brought forward was favourably entertained; and ultimately a joint company was formed, called the Paris and Rouen Railway Company: one condition stipulated by the English and their friends being, that they should appoint the engineer. This was readily acceded to; and the choice fell on Mr. Locke, who at that time had acquired a high reputation in England—one important element thereof being a confidence on the part of the public in the execution of his works within the estimates.

"Mr. Locke, on arriving in France to make the necessary arrangements, was impressed, from the information given to him as to cheapness of labour, with the idea that he should secure lower prices for the work than he had been in the habit of allowing in England: but the pretensions of French contractors seemed so much in excess of his expectations, that he suggested to the Board to invite

English contractors to come over and compete with those of the country. This suggestion was adopted; and several of the most prominent of the contractors in England were invited to examine the projects of the works, with a view to sending in tenders for their execution. Of these there were only three or four who really entertained the proposition. Mr. Brassey and the late Mr. William Mackenzie were of that number; in fact, I believe, they were the only two who thoroughly and seriously examined the matter. They soon discovered that the real contest would be only between themselves. The French contractors, owing, perhaps, to a want of practical knowledge of the execution of this new kind of work, to their limited resources of capital, and, still further, to the short time (as it appeared to them (allowed by the engineer for the execution of the line, framed their estimates on far too high a scale.

"Seeing this, Mr. Mackenzie and Mr. Brassey, not unwisely, agreed to join, and consequently tendered conjointly for the contracts as they came out, and succeeded, by competition, in securing the execution of (with a very trifling exception) the whole of the works. This was Mr. Brassey's *début* in France, and in fact the commencement of his practice in foreign countries.

"The works were commenced in 1841; and the line was opened to the public in May, 1843.

"Mr. Brassey fixed his residence on the line, and gave up the whole of his time and attention to it. In those early days this was, even to Mr. Brassey, a very heavy and important undertaking. Added to its extent, and the consequent and natural difficulties of organization and management, it possessed the new feature of being in a foreign country, where railway works were as yet unknown, and where, consequently, it was not easy to secure assistants in the shape of practical agents, foremen, and gangers, or even the necessary labourers, miners, and navvies accustomed to that style of work, and to the means of execution adopted by the contractor. All this considerably enhanced the difficulties, more especially as the whole time for completion was very limited, and necessitated, therefore, great energy, decision, and discernment in organizing rapidly a very large staff of *employés* of every description, and the bringing over from England numbers of workmen of all classes—amounting, at times, to several thousands." [1]

Railway construction has been one of the most gigantic series of enterprises of modern times, or of any times, and it seems to me that it cannot but be interesting to examine minutely how such enterprises were carried forward by one of the foremost men engaged in them.

Mr. Brassey took separate contracts for various portions of the line, being always able to underbid his foreign competitors, from the knowledge he had already acquired in railway-making, and especially from his having begun to collect around him a staff of well-tried and capable men. Throughout his life to form such a staff was one of Mr. Brassey's chief aims. He had, for this purpose, qualifications of the highest order. In the first place, he was skilful in his choice of men. Then he had a belief in the men he had chosen. If a man could not do well one thing that he had been put to do, he did not get rid of him, but would give him a trial in another branch of work. Mr. Brassey became well known throughout the labour market as an employer who was very loth to part with any man whom he had once employed.

He carried this practice to such an extent, that, in one or two rare instances, when his subordinates had opposed him, and even tried to go to law with him (but Mr. Brassey was a man very difficult to go to law with), he did not refuse to give these men further employment.[2] Moreover, he endeavoured so to regulate his work, that there should always, if possible, be employment for all his men, from the highest on the staff to the commonest labourer. This is a matter of great difficulty for the railway contractor. He is not in the position of a manufacturer, who may, even in times of distress, continue to employ his men, perhaps at a reduced rate of wages, but still employing them, heaping up goods for which there will be sure to come a demand at some time, But when there comes a slackness, or a dearth of railway enterprise, it is very difficult to continue to provide employment for all those persons who have been

---
[1] Mr. Murton's evidence.
[2] See Letter No. 2, in Appendix.

engaged in railway work during busy periods. The way in which Mr. Brassey managed, on these emergencies, was to subdivide the work he had to give into smaller portions; thus endeavouring to provide work for all his staff, for a given period, until better days should come round.

It will naturally be a matter of some interest to the reader to know how our fellow-countrymen, especially the navvies, got on in a foreign land. Mr. Brassey provided for them medical assistance and hospital accommodation, subscribing always very liberally to the hospitals that were in proximity to his men; and afterwards, with his usual generosity, often continuing those subscriptions when his men had left the country. There was much, however, of difficulty for the English in a strange country, which their employer could not provide against. They had to be employers as well as employed; and their mode of instructing the Frenchmen working under them, or working with them, was at first of a very original character. They pointed to the earth to be moved, or the wagon to be filled, said the word "d—n" emphatically, stamped their feet, and somehow or other their instructions, thus conveyed, were generally comprehended by the foreigner. This form of instruction was only applicable, however, to very simple cases, and some knowledge of the language had to be acquired by the men, for they could not afford to employ interpreters, as was done by persons of a higher grade in Mr. Brassey's employment. Several of these, and of their sons, soon acquired a competent knowledge of the languages of those countries in which they had such large negotiations to direct, and orders of all kinds to give. But among the navvies there grew up a language which could hardly be said to be either French or English; and which, in fact, must have resembled that strange compound language (Pigeon English) which is spoken at Hong Kong by the Chinese in their converse with British sailors and merchants. It must have had at least as much French in it as English, for it is stated in evidence that "the English learnt twice as much French as the Frenchmen learnt English." This composite language had its own forms and grammar; and it seems to have been made use of in other countries besides France; for afterwards there were young

D

Savoyards who became quite skilled in the use of this particular language, and who were employed as cheap interpreters between the sub-contractors and the native workmen. One of Mr. Brassey's agents, speaking on this subject, says:—

"It was not necessary to understand a word of English, but to understand the Englishman's Italian or French. That I found in many cases. A sharp youth, for example, would be always going about with a ganger, to listen to what he was saying, and to interpret to his (the youth's) countrymen."

It is pleasing to find that, after all, we have some power in the acquisition of languages, for several of these navvies did eventually acquire a considerable knowledge of French, not, of course, speaking it very grammatically, but still having acquired a greater knowledge of it, and a greater command of it, than they had of their native tongue.

On this railway between Paris and Rouen there were no fewer than eleven languages spoken on the works. The British spoke English; the Irish, Erse; the Highlanders, Gaelic; and the Welshmen, Welsh. Then there were French, Germans, Belgians, Dutch, Piedmontese, Spaniards, and Poles—all speaking their own languages. There was also one Portuguese, but he was a linguist in his way, and could speak some broken French.

This concourse of individuals, from various nations, took place wherever a railway was being constructed by English companies, in any part of the world which was not of a completely isolated character. It was therefore of great advantage that there should be some one language, such as that invented by the navvies, which should serve for the purpose of talk and instruction upon railways; and it is not surprising that this language should have been adopted wherever the English came in considerable numbers to be employed in the construction of a line, in any foreign country.

One of the first things to be arranged, was the lodging accommodation of the numerous bands of workmen, which frequently amounted to from ten to twenty thousand. Sometimes they were located in huts. On this particular railway (the Paris and Rouen line) there was no occasion for the construction of huts, for there are many villages, lying close to each other, all along the course of the river Seine. It is to be noticed that the Germans were content with much poorer accommodation than the other labourers. To use the words of an eye-witness, "They would put up with a barn, or anything."

Of the advantage which these railway works proved to the poorer inhabitants of those parts of the country through which these lines were carried, it is difficult to speak too highly. The "natives," as our Englishmen always called them, were provided with new tools, and learnt the use of them; were taught new forms of labour, and the benefit of organization in labouring; were paid regularly, and received a much higher rate of wages, sometimes double or treble that which they had been accustomed to earn. In making the railway from Charleroi to Givet, where the works were of a light character, Mr. Brassey sent out only a few Englishmen, to commence and superintend the construction of the line. One of the sub-contractors thus describes the effect upon the natives of the introduction of railway work. "When we went there, a native labourer was paid one shilling and three pence per day; but when we began to pay them two francs and two francs and a half a day, they thought we were angels from heaven." More provident and more abstemious than our countrymen, these natives contrived to make considerable savings; and they trooped back to their homes, often very distant, bringing not only sustenance and comfort to their wives and families, but having accumulated some capital for their own private enterprises at home. There are extensive districts in which the material prosperity of the inhabitants has been permanently raised by the savings which these hardy labourers realized, and brought back.

In 1843, the Rouen and Havre Railway, a continuation of the Paris and Rouen line, was projected to complete the communication between Paris and London, by way of Southampton.

"The works of the Havre railway were extraordinary in magnitude. The line, leaving the Valley of the Seine at Rouen, had to cross several important valleys to attain the plateau or summit level, and then to descend to the level of the port of Havre. This necessitated a large bridge over the Seine, many tunnels, eight or ten in number, several large viaducts of 100 feet in height, and huge cuttings and embankments; moreover, the whole of the work had to be completed in *two* years. Mr. Brassey took up his residence at Rouen, and laboured at this very heavy and important work with unbounded energy. I should say that, never up to that date, had such heavy works been carried out in so short a time. Although many of his people had had two years' experience in France, still, owing to the severe character of the work, there was much difficulty in obtaining the necessary labour, more especially as regards the mining, brickwork, and masonry. The contractors were again obliged to bring over from England hosts of bricklayers, from London or from any place where they could be found; and it may here be mentioned that, of all classes of railway labour, as a rule, the brickmakers and the bricklayers are the worst and the most unscrupulous, and great indeed was the trouble and expense they caused. The necessity also, of working night as well as day, rendered the supervision very difficult, particularly in the tunnels, and much anxiety was thereby occasioned to the engineers as well as to the contractors.

"During the progress of the works, a great accident occurred in the second section of the line, in the fall of the Barentin Viaduct—a huge brick construction of 100 feet in height and about one-third of a mile in length, having cost some £50,000; and which had, but a very short time previously, elicited the praise and admiration of the Minister of Public Works, and the other high French officials who visited it.

"This great downfall occurred a very short time before the proposed opening of the line. It is scarcely necessary here to seek to establish the causes of this failure; very rapid execution in very bad weather, and being built, in accordance with the contract, with mortar made of lime of the country (but with which the other smaller works had been successfully built), were no doubt the principal causes.

"Mr. Brassey was very greatly upset by this untoward

event; but he and his partner Mr. Mackenzie met the difficulty most manfully. 'The first thing to do,' as they said, 'is to build it up again,' and this they started most strenuously to do; not waiting, as many would have done, whether justly or unjustly, to settle, by litigation or otherwise, upon whom the responsibility and the expense should fall.

"Not a day was lost by them in the extraordinary efforts they had to make to secure millions of new bricks, and to provide hydraulic lime, which had to be brought from a distance. Suffice it to say that, by their indomitable energy and determination promptly to repair the evil, and by the skill of their agents, they succeeded in rebuilding this huge structure in less than *six months*.

"I should mention that, as one inducement to the contractors to open the Havre line a few months before the contract time, a premium of about £10,000 was offered them. This of course they stood to lose by this accident. The Company, however, in consideration of their marvellous and successful efforts to redeem the loss of time, allowed them the benefit of this sum, but the whole of the remainder of the expense they themselves bore. This is one of the many cases where, in spite of all loss, of all difficulty, that determination never to shrink, upon any pretext, from a contract, fully evinced itself; and therefore, it is a case worthy of note.

"Allusion may appropriately be made here to Mr. Brassey's personal management of works; for, at this period of his career, he had not, as at a later time, multifarious contracts in hand in different countries. He was therefore enabled to give up nearly all his time to the works in France; he, consequently, gave them his direct personal management, being assisted mainly by resident agents, each having the superintendence of a district of a few miles." [1]

Mr. Harrison, at whose house Mr. Brassey was at the time he received the news of the Barentin accident, says that the only remark he made, was simply, "I must leave you," and that he at once sent for Henry Chambers, who

---

[1] Mr. Murton's evidence.

had charge of the bricklaying at the viaduct, to give instructions about the rebuilding.

I have endeavoured to give somewhat of a survey of what were the elements, moral and material, which went to form a great railway enterprise in a foreign country. If we look at the several persons and classes engaged they may be enumerated thus:—There were the engineers of the company or of the government who were the promoters of the line. There were the principal contractors, whose work had to satisfy these engineers; and there were the agents of the contractors to whom were apportioned certain lengths of the line.

These agents had the duties in some respects of a commissary-general in an army; and, for the work to go on well, it was necessary that they should be men of much intelligence and force of character. Then there were the various artisans, such as bricklayers and masons, whose work, of course, was principally that of constructing the culverts, bridges, stations, tunnels, and viaducts—to which points of the work the attention of the agents had to be carefully directed. Again, there were the sub-contractors, whose duties I have enumerated: and under these were the gangers, the corporals, as it were, in this great army, being the persons who had the control of small bodies of the workmen, say twenty or more. Then came the great body of navvies—the privates of the army, upon whose endurance and valour so much depended.

It remains only now to imagine all these numerous bodies in full and harmonious action. To take this out of the field of imagination, and to give a real description of the scene, I cannot do better than quote the words of one of Mr. Brassey's time-keepers, from whose evidence I have gained much.

"I think as fine a spectacle as any man could witness, who is accustomed to look at work, is to see a cutting in full operation, with about twenty wagons being filled, every man at his post, and every man with his shirt open, working in the heat of the day, the gangers looking about, and everything going like clockwork. Such an exhibition of physical power attracted many French gentlemen, who came on to the cuttings at Paris and Rouen, and looking at these English workmen with astonishment, said 'Mon Dieu! les Anglais, comme ils travaillent!' Another thing that called forth remark, was the complete silence that prevailed amongst the men. It was a fine sight to see the Englishmen that were there, with their muscular arms and hands hairy and brown."

CHAPTER V.

LABOURERS OF DIFFERENT COUNTRIES.

Foreign workmen.—Origin of the word " navvy."—The navvy's conduct abroad.—Foreign tools.—Plate-layers.—The miner.—The engineer.—The Piedmontese.—The Neapolitans.—The Germans.—The Belgians.—Tipping.—Belgian railway-making.—Belgian system of wages.—Mr. Hawkshaw on labour.

THERE is not anything which is more significant of a man's nature than his mode of working. Work is the outcome of the whole man. The same remark may be made of nations as of individuals; and, throughout the world the different sections of it work very differently.

Intellectually speaking, this difference is very manifest, and it is to be noted throughout the literature of the various nations that have any literature at all. Some peoples are habitually accurate, and delight in neatness and in finish; others go about their work in a somewhat slovenly and unprepared manner, but aim at larger though less complete results. The individuals of some races can work independently, and so do their work best: others need a large amount of direction and supervision. Nor are these differences less visible in manual labour. A man such as Mr. Brassey, having to execute great works in foreign countries, had, of necessity, to pay great attention to these differences in the capacities of the various people by whom his work was to be executed; and he had to apportion their labours and regulate his payments accordingly.

The first promoters of railways in England had one considerable advantage as regards a certain class of labourers who were at that time ready to their hands. The general subject of locomotion of all kinds, had for a long time attracted great attention in England. Road-making, as I have said before, prospered to a greater extent in England than in any other country: but it was not from the makers of roads that the contractors for railways drew their best supplies of labour of the lowest, but not the least important, kind. The men who did the hardest work in railway making, were those who had been engaged in a similar kind of work, requiring cuttings and embankments,— namely, in the formation of canals. Hence the name of "navigator," which was soon abbreviated into that of "navvy."

These men, having been employed in the construction of canals, were eminently fitted for railway making. Indeed, the work to which they had been accustomed was such as required, in some respects, even more care and attention than railway work; for the best of brickwork and masonry, and well-made earth-works, were necessary to make a canal secure. Those persons, who had to *direct* the commencement of railway making, had far more difficult problems set before them than had been encountered by the constructors of canals; but, for much of the inferior work, the common labourer at canal making had received a training which more than fitted him for his share of the work on railways.

The English navvy is generally, in the first instance, an agricultural labourer. He is, however, but an indifferent specimen of a labourer when he first commences, and he earns only about two shillings a day. Gradually he acquires some of the skill of his fellow-workmen; and then he rises into a higher class, receiving three shillings a day. Ultimately, if he is a handy man, his work becomes worth still more, and his wages will rise to four shillings a day.

Mr. Ballard's evidence with respect to the amount of labour done by the English navvies is very precise, and very valuable. He states, as his opinion, that "the labour which a navvy performs exceeds in severity almost any other description of work." He says that "a full day's work consists of fourteen sets a day." A "set" is a number of wagons—in fact, a train. There are two men to a wagon. If the wagon goes out fourteen times, each man has to fill seven wagons in the course of the day. Each wagon contains two and a quarter cubic yards. The result is, that each man has to lift nearly twenty tons weight of earth on a shovel over his head into a wagon. The height of the lifting is about six feet. This is taking it at fourteen sets a day; but the navvies sometimes contrive to get through sixteen sets, and there are some men who will accomplish that astonishing quantity of work by three or four o'clock in the afternoon—a result, I believe, which is not nearly equalled by the workmen of any other country in the world."

There are no trades unions amongst the navvies, and there were very seldom any strikes. This statement applies not only to the labourers who worked on the Paris and Rouen Railway, but generally to those employed throughout the course of all Mr. Brassey's railway undertakings.

With regard to the conduct of the navvies, when off duty, in France they were at first rather troublesome. Brandy was cheap, and they had unfortunately a tendency to drink it freely, which was not the French habit. Mr. Mackay, after admitting this tendency to drink on the part of his countrymen, goes on to say:—

But after a short time the French found that they were a good-natured sort of people, who spent their money freely. Hence they were always kindly received; and even the gendarmes themselves began very soon to see which was the best way of managing the Englishmen. They got sometimes unruly on pay-day, but not as a rule.

The English navvy came to his foreign work without wife or family. After he had been employed upon one railway abroad, he generally sought for similar employment; or, if he was an intelligent man, aimed at higher employment on other railways in foreign countries. The unmarried Englishmen frequently married foreign wives; the married men, who had left their families behind them, sent home money periodically to their wives; and in either case they often sent money to their parents.

The navvy, like most of his fellow-countrymen, of whatever rank or occupation, scorned to adopt the habits or the dress of the people he lived amongst. An accurate observer thus bears witness to this fact:—" I never found a navvy adopt any other costume, but the English navvy's costume. I have seen him generally with a piece of string tied round his leg below the knee, and with high-low boots laced up, if he could but get them made."

Mr. Brassey, on commencing his work in France, must have had to consider with great care the comparative merits and powers of the English and French labourer.

It was found that the tools in use abroad were of a most inferior description. The French used wooden spades. Their barrow was of a bad form, and they had very inferior pickaxes. These defects could easily be remedied, but not so the manner of carrying out the work, and the men's small power of working. This was such, that their work was found to be worth only two francs a day, while the English labourer would earn four francs and a half. In time, however, the Frenchman living better and learning more, his work became worth four francs a day; and, gradually, in any work undertaken by Mr. Brassey in France, the number of the English labourers was lessened, and the number of the French labourers increased, until, at last, the great bulk of the railway work in that country was done by Frenchmen.

There is one fact connected with the payment of wages which deserves, I think, to be noted. The Frenchman, as I said, received two francs a-day: the Englishman four francs and a-half. Now, the respective results of their work was not exactly in this proportion, for in the mere moving of earth it was found that the Frenchman was able to "shift," as they call it, half as much material as the Englishman. It seems to me probable that this disproportion in favour of the Englishman is an indication of the value attached by the contractor to any additional speed in the execution of his work, and for extra rapidity of execution under pressure more reliance could be placed upon the Englishman than the Frenchman. It is observable, throughout Mr. Brassey's career, that he attached great value to the rapid execution of any work he had undertaken; and if any disaster occurred, his first thought seems to have been, not who was to blame, or upon whom the loss should fall, but how the work in question should most promptly be restored. The other matters were to be afterthoughts, and were always dealt with by him in the most liberal manner; for the proverb that "It is no use crying over spilt milk," was often in his mouth, and was acted up to as well as quoted.

As an instance of what I have just said, Mr. Henry Harrison, Mr. Brassey's brother-in-law, says, "I may mention that at Rugby, during the execution of the Trent Valley contract, there was a great difficulty as to bricks; the clay was very unfavourable for making them, and, after a considerable sum of money had been spent in making bricks of an inferior quality for the railway, it was found necessary to abandon the attempt, and to provide bricks from another source. I naturally felt greatly discouraged at the loss which the contractor had sustained; but Mr. Brassey observing this, encouraged me not to be unduly depressed, saying that I must never take such troubles to my pillow; as the loss of bricks was a matter of secondary importance so long as the line was completed within the stipulated time. This encouraging bearing on his part, in this and many other similar difficulties, was a great support to members of his staff, and often cheered them in their work amidst circumstances of great discouragement."

But to resume the description of the contrast between the foreign labourer and the English. There were certain branches of the work which were specialities for them, "plate-laying," for instance, was originally, and remained for some time, an English speciality. Ultimately, however, the Frenchman acquired the art of plate-laying, and he does it now exceedingly well.

There is one branch of work in which the English labourer has always been pre-eminent. I refer to that of a miner. It requires special energy and endurance, as the conditions under which a man has to labour are exceedingly unfavourable. His clothes are frequently saturated with water, and he has to breathe in a most oppressive atmosphere. Nor is this all. It requires very considerable courage to undertake the risks involved in that branch of mining which consists in the construction of tunnels. I adduce here the evidence of Mr. Charles Jones, who has had great experience in this kind of construction. He says:—

The workmen have to labour in a space which is temporarily shored up by timber, and the pressure of the earth is constantly putting the timber structure to a great strain.

At times you hear alarming creaking noises round you, the earth threatening to come in and overwhelm the labourers.

On being asked whether, under these most trying circumstances, the peculiar national virtues of the English labourer are not specially manifested, his reply was:—

Yes; it is often necessary to strengthen the temporary timber structures by adding additional beams, or placing uprights underneath the planks overhead, which are yielding to the weight above them. It requires a considerable amount of courage in the men employed in this kind of work. If they shrink from facing a certain amount of danger, the whole structure would sometimes come in upon them, thereby endangering their lives, and retarding very considerably the progress of the works.

It is a matter of some interest to observe what differences there are between the higher classes of Frenchmen and Englishmen employed in railways—for example, the engineers. Here, as might be expected, the difference of the national character was very visible. The English engineer had, to use a common phrase, more practical "go-a-headism" in him; but then it must be taken into consideration that the French engineer partook of the nature of his Government, and had a very different aim from that of the Englishman, following therein the views of his Government, which aimed at making everything most durable. Whatever he superintended, whether it was a bridge or a viaduct, was not merely to last for ninety-nine years, but at the end of that time was to be as durable as ever.

The different policy of the English and French Governments, arising from the different natures of the two nations, is remarkable. The English Government, of which, soon after the commencement of railway-making, Sir Robert Peel was at the head, admitted to the uttermost, or nearly to the uttermost, the principle of free competition in regard to the formation of railways in England.

I proceed now to give an account of the processes of working of some of the other foreign workmen, amongst whom Mr. Brassey brought his bands of English workmen as pioneers in the art of railway-making.

Their tools were mostly of a similar nature to those of the French, which were only "Fly-tools" as one of Mr. Brassey's staff graphically describes them. In the work on Italian railways, great difference was found as regards the character, and mode of working, of the different races who now constitute the Italian kingdom. The Piedmontese were found to be very good hands. Indeed, one of Mr. Brassey's agents, Mr. Jones, thus expresses his opinion of their merits:—" For cutting rock, the right man is a Piedmontese. He will do the work cheaper than an English miner. He is hardy, vigorous, and a stout mountaineer; he lives well, and his muscular development is good."

Speaking of the Piedmontese generally, he says:— "They are quiet, orderly men; they are not often tipsy or riotous; and they go to their work, and do it steadily, putting by money before they go back to the hills."

He then proceeds to make the following general observation:—" It is found that all the people born in the mountains, and on poor lands, have more virtue than those who are born in the plains, and in luxurious places."

With regard to the Neapolitans, they would come in large troops to the places where railways were being made; and these troops were under the command of certain chieftains, as it were. The leading men, each of them followed by about a thousand labourers, would take a considerable length of earthworks to execute—for instance, ten miles of light earthworks, and side-cuttings. *But they would not take any heavy work.*

The labourers brought with them their fathers and grandfathers and their male children; but they left their women behind, in their native villages. Then they build huts of mud and trees, which, during the day, were left in charge of the old men, who also undertook the cooking.

On account of the climate (I am specially alluding to work in the Maremma), these bands of men could not work more than six months at a time. They usually earned a franc a day; but sometimes at piece-work, they made very nearly two francs a day, working fifteen or sixteen hours; but their work was chiefly that of removing light soil in baskets. As the witness well said, "English navvies would not understand this way of doing work." These poor Neapolitans, who, by the way, chiefly came from the South of Naples, near the Abruzzi, were a very frugal and temperate set of men. They ate bread and vegetables, and drank only water. They had also tobacco, a little coffee, and a small quantity of goats' meat now and then. After their six months' work was done, they would return to the mountains, "with the old men and little boys, carrying the kettles and pans, and taking their money home with them."

The men from Lucca were a race, who, for working powers, might be placed between the Piedmontese and the Neapolitans, being less hardy than the former, and more so than the latter.

With regard to the Germans, it is worthy of remark, that in the opinion of a man who had great opportunities of judging, they had less endurance than the French. His words are:—"I have seen the Frenchmen 'harry,' that is, overcome their work, and distress the Germans in the power of endurance. And from that circumstance, during the last war, knowing Frenchmen very well, I felt almost certain that they would win the day."

It must be noticed, however, that the Germans employed by Mr. Brassey on the Paris and Rouen Railway were chiefly Bavarians.

We now come to the Belgian workmen. They were, as might be expected, good labourers; and they had had some experience in railway-making before Mr. Brassey's men came among them, for the King of the Belgians had already taken up railway-making. But they, like the rest of the world, were greatly behind the English in several of the processes of work. I will give a remarkable instance of their backwardness, which, moreover, is of general application.

The English, very early in their career of railway construction, bethought them of the excellent idea of filling up the hollows from the heights—without any intermediate operation. They did not remove the earth which they took out from a height to form a cutting, to what is called a "spoil bank," except in particular cases. They aimed at bringing the earth, which they were obliged to take from the height in order to reduce it to the proper level, at once to the hollow which they had to fill up. This they did in a most ingenious manner. As they formed their line, they laid down temporary rails upon it; then they filled a number of wagons with the earth from the cutting; and these wagons, when filled, were drawn by horses out of the "cutting" to a certain point near the end of the embankment." There, the wagons were detached from each other. Afterwards they were attached singly to the "tip-horse," who would trot or gallop with them nearly to the brink of the bank, where the horse being set free by a peculiar contrivance, would step on one side, and the wagon, running on by its own impetus, and coming against a sleeper placed at the end of the rails, shot out the earth into the proper place. This process is one alike of much skill and some daring; and the idea of it is very creditable to the inventors. Anybody who has watched the whole proceeding, must have been greatly interested by it. Indeed, one lingers on, watching the process with satisfaction. A very forcible phrase is in use among the the navvies, when they propose to send a large number of wagons heavily laden from the cutting to the spot where the wagons are to be charged. They say, "we'll run 'em in a *red* un ; " "red" standing in their language for large—a phrase which fully corresponds with other forms of their language.

I will now give a description of how this process affected the Belgians, when they first saw it; and I give it in the words of Mr. John Mackay, one of Mr. Brassey's sub-contractors :—

The Belgians had never seen wagons before like ours. I began laying down a temporary road : they could understand that, but when I began to make a "turn-out" for a "lay-by," for empty wagons to come into as the full wagons passed, they could not make that out at all. They assisted me, but I was obliged to manage this work myself. At last I got the wagons on the road and began to fill them. I got the "tip" harness on to the horses, but had no one to drive, and so I was obliged to drive myself. I then selected a nimble Belgian for the purpose of teaching him to become a driver, but I had to drive a horse myself for about two days. I gave this man a horse, and told him to start the first "set" of wagons—he looked round the horse, then at the spring bar that the chain was hooked to, but could not make it out. However, I got the first set of wagons filled, and, being driver, I drew the wagons out of the cutting to the tip; but, to my great surprise, when I looked back, I saw that every man in the cutting was following me. They ranged themselves on each side of the bank until they saw the first set of wagons tipped. They could not make out how I pulled the horse out of the road; or how the horse escaped, and the wagon went ahead. This went on for three or four sets, until they understood the mode of operation.

Before Mr. Brassey's agents went out, that comparatively humble instrument, the barrow, had been exclusively used by the Belgians in railway works. If there was a cutting and embankment contiguous, they took a small part of the earth from the cutting for part of the embankment, but all the rest they put out to "spoil." Then they made up the embankment by barrow-loads of earth from "spoil banks." I need hardly say that the English wagon, and the mode of using it just described, have now been generally adopted throughout the Continent.

I cannot here refrain from giving another remarkable instance of the character and ingenuity of the English, which although not manifested in railway work, was carried into effect by the English navvy. It occurred on the works of the Paris fortifications commenced at that time in Louis Philippe's reign, at the instance of M. Thiers.

When the Paris and Rouen Railway was completed, some of the English navvies went up to Paris and tried to contract with the Government engineer to get some of the work to be done by "butty-gangs," and they obtained some work at French prices.

There was a "fosse" or ditch all round the works, and the stuff from the ditch was taken out to be put behind the masonry. The Frenchmen used zigzag roads, to get up which was almost interminable, and all the stuff was taken by them out of the ditch by barrows, and wheeled up by this zigzag arrangement on planks, until they got to the top of the wall. When the navvies began working they at once put a pulley upon the top of the wall, to make a "swing run" with a rope, so as to be able to swing the rope, and the empty barrow went down to the bottom by its own gravity; then they had horses on the top, which pulled the loaded barrows vertically up. They worked that way for about six weeks, and earned fifteen francs a day each. Then, no doubt, the French engineers began to see that these Englishmen were making them pay an enormous sum of money for the work, and they reduced the price; and the Englishmen, not having anywhere else to go, had to submit. When the next pay-day came round there was another reduction made to five francs a day. Then the Englishmen began to slacken speed, and complain that their employers cheated them in the measurement and in their pay—they would not do more than five francs' worth of work a day, and took it easy.

E

To revert to the Belgians. There was one remarkable point to be noticed about the Belgian workmen, at least about those who were employed on the Sambre and Meuse Railway. Money not being so easy, when easy at all, in any country as in England, the English contractor, or subcontractor, invariably paid his men with great punctuality; a mode of payment which it had not always been in the power of the native contractor to adopt. When Mr. Brassey took the Sambre and Meuse Railway contract, his agent being always regularly supplied by him, introduced the English custom of paying the men every fortnight. After this had gone on for some little time, the Belgian labourers, in a body, petitioned to be paid monthly. This naturally rather astonished the contractors; and one of them gives an account of the whole proceeding. He assembled his men together, and asked them what reason they had for making such a request. They replied that they would " rather have their money once a month, because they would have more to take at a time." "But," said he, "you will want some 'sub,'[1] during the month, and it will be quite as inconvenient to 'sub' you as to pay." "No," they said, "they would not want any 'sub.'" He then asked them how they were paid by their own contractors. They replied, "every six weeks, sometimes every three months, and sometimes we get none at all." I believe their request was listened to. The truth was, as the sub-contractor remarks, "that they had implicit confidence in Englishmen, and they thought they could lay out their money better when they had a lump-sum at once."

Mr. Hawkshaw says, speaking on the relative value of unskilled labour in different countries, "I have arrived at the conclusion that its cost is much the same in all. I have had personal experience in South America, in Russia, and in Holland, as well as in my own country; and as consulting engineer to some of the Indian and other foreign railways, I am pretty well acquainted with the value of Hindoo and other labour; and though an English labourer will do a larger amount of work than a

---

[1] A payment on account for " subsistence."

Creole or Hindoo, yet you have to pay them proportionately higher wages. Dutch labourers are, I think, as good as English, or nearly so; and Russian workmen are docile and easily taught, and readily adopt every method shown to them to be better than their own." [1]

[1] See Letter No. 7, in Appendix.

CHAPTER VI.

MR. BRASSEY'S SKILL IN DEALING WITH SCHEMES AND
CALCULATIONS.

Skill in estimating.—Cost of railways at home and abroad.—Buckhorn
Western tunnel.—Mental calculation.—General accuracy.

WE have seen how a railway is started; how the enterprise is manned; and we have gained some notion of how the work goes on when it is in full operation. This, however, gives but an inadequate representation of the labours of such men as Mr. Brassey, and the agents employed under him. A large part, both of his work and their work, consisted in making calculations respecting the different schemes which were put before him; many of which, as might be expected, after much thought and labour had been given to them, had to be rejected, or came to nothing, at least for him. In fact, I am told that Mr. Brassey unsuccessfully tendered for works to the amount of nearly £150,000,000. In speaking of this investigation of schemes, one of Mr. Brassey's agents observes, "this forms a most interesting part of our experience, as it has occupied almost one half of our lives."

There was no part of Mr. Brassey's work in which he showed more ability than in making these investigations, and coming to a conclusion upon them. Here, too, his sagacity in the original choice of agents, and in his subsequent mode of dealing with them, came into full play. Moreover, his great powers of calculation, for, though not claiming any of the wondrous powers of a Bidder, Mr. Brassey held a high place among mental arithmeticians, found a fitting field for exertion.

When any scheme was submitted to Mr. Brassey, he was accustomed to deal with it in this manner. The proposal was generally accompanied by plans and sections of some kind, and a few calculations. Mr. Brassey then sent for one of his agents, in whom he had confidence, and who would probably be one of the persons entrusted to carry out the project if it were found to be acceptable. Mr. Brassey handed the papers to this agent, and asked him in general terms to look at them, and in the first place to say whether the figures appeared to him to be anything like what they ought to be, or what the work might be done for.

The report of the agent was generally a verbal one, not occupying, perhaps, more than five minutes. If the report was such as to induce Mr. Brassey to suppose that there might be something in the scheme, he would say to the agent, "Then you had better go off and see to it."

Mr. Brassey would always have the project considered by some of his own people on the ground; and one of the points of investigation which he insisted upon being made with great accuracy, was the following:—namely, whether the length of line stated in the project was nearly the same as the length of the line would really be. His reason for causing this part of the enquiry to be made with care, was one which gives a good instance of his foresight, and shrewdness, especially as applicable to foreign railways. Very often on the length of the line depended the value of the Government guarantee to the shareholders. If the line proved to be much longer than was supposed, the amount of money at first thought to be sufficient would not prove to be sufficient. Then subsequent loans would have to be obtained; and Mr. Brassey's experience showed him that these subsequent loans were almost always obtained on unfavourable terms.

The agent carefully ascertained the length of the proposed line; examined the sections; went into the price of material and the price of labour in the country. Finally, he "took out the quantities," using the plans and sections supplied by the promoters. The agent then came back to give a report to Mr. Brassey.

When a line had to pass through easy country, the agent's labour in making the report, and Mr. Brassey's in coming to a conclusion upon that report, were but light. In a mountainous country, however, the labour was of a very different kind, and the most minute investigation of details was necessary.

As an instance of the manner in which Mr. Brassey dealt with those whom he employed, I may adduce the evidence of one of his agents. He had gone for Mr. Brassey to investigate and report upon a proposed railway in Holland. With regard to this railway, Mr. Brassey was required to sign a preliminary contract, which involved on his part such a knowledge of all the circumstances as would enable him to frame his estimates. The agent says:—

> The investigation which I had made was his only source of information. He said, " Let me see how you have got at that." I came to his office, and it took us a quarter of an hour at least, to look over details, which was a great deal of time for Mr. Brassey to give to any one matter.
>
> The length of the proposed line was about 120 kilomètres, and the estimated cost was about twenty-seven millions of francs. I had considered it for a long time; and when he went into the matter, in a quarter of an hour, or twenty minutes at the outside, he examined all the details. I mean to say, that in that short time he turned them all over, and stopped at the difficult points; and in the case of one of the bridge estimates, he went through my figures minutely, to see whether I had included all the details for the foundations, and had, in fact, included every detail which was likely to be involved in the execution of the work. He looked at the details for culverts, to see how I got at their price, and then investigated the calculations as to the price of the brick-work, of the stone-work, and the average quantities of earthwork. I had worked them out, to see what the average was upon separate lengths of the line. At the end of this investigation, he said, "That will do." Never after that did he look into any estimate of mine in such detail.

At the same time that we dwell upon the apparently wonderful way in which Mr. Brassey, and other great employers of labour, come to these conclusions, so rapidly and yet at the same time so judiciously, we must recollect that the results of the experience of these men often assume a very distinct form. For example, in railway construction, large experience seems to show, as was pointed out in the last chapter, that there is very little difference between the cost of labour in this country and in foreign countries: at any rate, not such difference as cannot easily be allowed for.

Of the enormous difference that exists in the expenses attendant on the construction of railways, and of the consequent care that must be taken in forming estimates, I am enabled to give a very remarkable illustration. In the tunnel that was made on the Salisbury and Yeovil Railway in 1860, the difference of expense was as follows:—The construction of a yard at one end cost £12 while a yard at the other end cost as much as £120.[1]

In reverting to Mr. Brassey's skill as an arithmetician, I may make a remark, which I think may be of some service in the way of education. I think it may be observed that all the men who have great powers of mental calculation, or who, as lawyers, statesmen, men of business, or authors, show great readiness in speech or action, or vast memory as regards facts, have made it a rule, or at any rate a practice, to rely upon that memory alone, and have not indulged in much writing of notes, to subserve the memory. If I am right in this statement, and I have many memorable examples, such as the late Lord Lyndhurst and the late Baron Rothschild, to adduce in my favour, it would indicate that in youth great pains should be taken to cultivate the services of that most admirable friend and servant, the memory.

I am particular in insisting upon this point in reference to Mr. Brassey. I am going to mention a striking instance of his powers of calculation; but before doing so, I should like to show how he dealt habitually with his memory, and what trust he put in it. He never condescended to that weak practice of making many notes. He had dealings with hundreds, I may almost say with thousands, of people. They came and told him their views and their wishes, their schemes, their intentions, and their grievances. He heard them all;[2] and if he did not reply to them at the time, as it was often impossible to do, for he had to make enquiries in relation to what they stated, it was perfectly certain that they would receive answers in writing, showing a complete knowledge of all the facts which it was necessary for him to refer to. He was one of the greatest letter-writers ever known. Retaining in his mind all he had to write about, he was ready at any halt in his innumerable journeys, if it were only a halt of a quarter of an hour at a railway station, to sit

[1] Mr. Harrison's evidence.
[2] See Letter No. 5, in Appendix.

down and write several letters, generally of the clearest and most distinct nature, embodying all the requisite facts and circumstances. This masterful memory would never have been attained if Mr. Brassey had been one of those persons who weakly, as I think it, rely upon notes, or written materials of any kind, in any matter where memory alone may be expected and encouraged to do the work.

I now give the instance of his power of mental calculation, which I have before alluded to. I take the words of one of his staff:—

> After dinner Mr. Brassey, and Mr. Strapp, the principal representative of the contract, were going into details of prices; and Mr. Brassey called me to take a chair by them, and go into details of prices of different kinds of work. I noticed especially, after we had given him the cost, for instance of a bridge—all the details of the bridge, and the total cost—he said; "How many bridges of that kind are there upon the line at the same prices?" Again, as to the culverts, or bridges of a different size. After going through the quantities of the masonry, we went into the earthworks, and talked about the nature of the material and average length of "lead" from the different cuttings, and how much would this cost, and how much such and such a bank would cost, or such and such a deviation, and the prices of different parts of the works. Then we came to the question of rails :—they would cost so much delivered at the station, and so much delivered on and along the line. There was a very great distance to convey them: but there was a great facility for carting during the winter months, owing to a great deal of wood being carted into the towns for the winter fires, so that we could get the rails carted by a sort of back carriage on sledges.
>
> We had to get all these details as to carting the rails, and we gave him the average cost of the rails on the line; and then, almost in a few seconds, he arrived at the approximate cost of the line per mile, mentally.

Before ending this chapter, I should observe that Mr. Brassey's conclusions were hardly ever found to be wrong as regards the subject-matter upon which they were employed, when that subject-matter was the construction of a railway. Whenever he fell into any difficulty, it was generally the result of some financial embarrassment; and over such embarrassments no man can be said to have a complete control. They depend upon large general causes, against the operation of which a prudent man may sometimes guard; but by which even the most prudent of men, engaged in large commercial undertakings, are liable to be surprised.

CHAPTER VII.

MR. BRASSEY'S WORK BECOMES MORE EXTENSIVE.

(A.D. 1845.)

Attention to details.—Visits of inspection.—Would not enter Parliament.—The Emperor of the French.—Mr. Brassey's activity.

MR. BRASSEY now entered into what may be called a new phase of life. He was no longer the contractor for one or two lines only, but had become a man to whom many persons resorted when they were the projectors of great railway schemes. This is the kind of change which occurs to most men, of any note, in the course of their career. They begin by doing well some one or two things in which they are totally engaged, being obliged to attend to every detail and to make themselves thoroughly masters of the work. They do not so much superintend as execute.

It is sometimes found, however, that this change is not attended by fortunate results, and that the man who can govern and direct a few persons under him—those persons remaining constantly under his eye—is perfectly incompetent to manage great undertakings in which men, who have performed functions exactly similar to his own, of which he is perfectly cognizant, are to be mere units in these his greater, or at least more extensive undertakings. The error which those commit who fail when they are obliged, in the course of their career, to enter upon a much larger field of operations, is, that they still pay too much attention to details. It is not every man who is fitted to be promoted from an inferior rank, however well filled by him, to one of generalship and wide command.

But certainly Mr. Brassey was found to be equal to the change of position and of duties which he had now to undertake. It is almost curious to observe, from the evidence of all those who acted under him, how he was henceforward wont to look to results rather than to minute details. As his works extended—and in 1845 he had on his hands, no less than thirteen heavy contracts which alone represented a length of about 800 miles, scattered in various parts of Scotland, Wales, France, and England—he evidently took the greatest care not to waste his time, thought, or energy, upon any work of detail which he could get as well done for him by others. I always thought, what an admirable official man he would have made; for skill in high office greatly consists in discerning and making use of the powers and intelligence of subordinates, and of reserving all that force, of which no man has much to spare, for the solution of intricate questions and the determination of high resolves. The man who is oppressed by details will never be able to do this. The man who has not in his own time dealt much with details, and learnt to master them, also will not be able to do this: and moreover, without having previously mastered details, he will never acquire that respect from his subordinates which he ought to command; for they believe intensely in the skill which their superior may be able to show in dealing with details.

Henceforth Mr. Brassey was not to be found doing any of the work which an agent or sub-contractor could do just as well. His visits to his various works necessarily became more infrequent. Indeed, one of his assistants remarks that if Mr. Brassey frequently visited any line, it was a proof that there was some difficulty there, or that something was going wrong. His visits were still looked forward to as occasions for great rejoicing, as opportunities for the redress of grievances, and for the settlement of all questions of difficulty. He became, as it were, the great consulting physician in railway matters, only making his appearance on critical occasions.

It must not be supposed, however, that he began to work less; but only that his work, being of a different nature, did not demand the minute supervision which he had been accustomed to give to it. To use the words of one of his agents, "Mr. Brassey never ran away from the duties of his calling. He was a contractor for public works, and did not pretend to be anything else." He has been heard to

MAP OF ENGLAND, SHOWING RAILWAYS CONSTRUCTED
BY MR. BRASSEY.

say, "It requires a special education to be idle, or to employ the twenty-four hours in a rational way, without any particular calling or occupation. To live the life of a gentleman," he would add, "one must have been brought up to it. It is impossible for a man, who has been engaged in business pursuits the greater part of his life, to retire: if he does so, he soon discovers that he has made a great mistake. I shall not retire: but if, for some good reason, I should be obliged to do so, it would be to a farm. There I should bring up stock, which I would cause to be weighed every day, ascertaining at the same time their daily cost, as against the increasing weight. I should then know when to sell, and start again with another lot."

One day when Mr. Brassey and Mr. McClean were dining with Mr. Giles, the conversation turned upon the immense fortunes of several noblemen, and how men, born to £50,000 a year and upwards, spent their income. Upon which Mr. Brassey said, "I understand it is easy and natural enough for those who are born and brought up to it, to spend £50,000 or even £150,000 a year; but I should be very sorry to have to undergo the fatigue of even spending £30,000 a year. I believe such a job as that would drive me mad."

Neither would Mr. Brassey suffer himself to be allured from the labours which were most congenial to him, and in which he felt he could do most service to the world, by any employment which some people might consider to be of a higher kind.[1] He respectfully but steadfastly refused all offers to bring him into Parliament. His ambition, as has been already stated, was to be the great contractor, to furnish large and continuous employment to his fellow-countrymen, and to the "natives" of other countries. He felt this to be a sufficient object of endeavour, and one that would occupy him throughout his lifetime, as long as his power of promoting so great an object remained to him.

It cannot be said, either, that Mr. Brassey was much attracted by honours of any kind. One of his agents thus describes his reception of the Cross of the Iron Crown from the Emperor of Austria:—

---

[1] See Letter No. 3, in Appendix.

Returning from Vienna, Mr. Brassey was waited upon at Meurice's Hotel, Paris, by one of his agents, who arrived in the room at the very moment his travelling servant Isidore was arranging in a little box the Cross of the Iron Crown, which Mr. Brassey had just before received from the Emperor of Austria. Made acquainted with the circumstance, the agent complimented his chief as to the well-merited recognition of his services, &c., and the conversation continued on Foreign Orders generally. Mr. Brassey remarked that, as an Englishman, he did not know what good Crosses were to him; but that he could well imagine how eagerly they were sought after by the subjects of those Governments which gave away Orders in reward for civil services rendered to the State, &c. He added, that in regard to the Cross of the Iron Crown, it had been graciously offered to him by the Emperor of Austria, and there was no alternative but to accept this mark of the Sovereign's appreciation of the part he had taken in the construction of public works, however unworthy he was of such a distinction. "Have I not other Crosses?" said Mr. Brassey. "Yes," said his agent; "I know of two others, the Legion of Honour of France, and the Chevaliership of Italy. Where are they?" But as this question could not be answered, it was settled that two duplicate crosses should be procured at once (the originals having been mislaid) in order that Mr. Brassey might take them across to Lowndes Square the same evening. "Mrs. Brassey will be glad to possess all these Crosses."

This simple remark of Mr. Brassey's contained a world of meaning, for although he attached no importance to such matters himself, he knew and appreciated the great interest taken by his wife in all things touching his fame.

I may here remark that when Mr. Brassey received the Cross of the "Legion of Honour" from the Emperor of the French, the Emperor invited him to dinner at the Tuileries; and at this dinner Mr. Brassey sat near the Empress, with whose grace and manner he was much charmed, and was especially pleased with her kindness in talking English to him during the greater part of the time.

"Although the great bulk of the enterprises achieved by Mr. Brassey was the formation of railways, with all their appurtenances, still his operations were by no means limited to this branch; he interested himself and took prominent part in various other vast projects, such as steamships, mines, engine factories, marine telegraphs, and in many cases he became the main support and the largest proprietor of many of these costly undertakings.

"Mr. Brassey may therefore be said to have been a man of almost universal utility to the civilized world, and the results of his long industrious career have not been merely the accumulation of a colossal fortune for himself and those to follow him, but have had a far more important character, great services rendered to mankind; and as such, his memory cannot fail to be regarded with pride not only by all who knew him, but by all his countrymen and the world in general."[1]

Mr. Brassey's personal activity is well described by Mr. Henry Harrison, his brother-in-law. He says, "I have known him come direct from France to Rugby. Having left Havre the night before, he would have been engaged in the office in London the whole day; he would then come down to Rugby by the mail train at twelve o'clock, and it was his common practice to be on the works by six o'clock the next morning. He would frequently walk from Rugby to Nuneaton, a distance of sixteen miles. Having arrived at Nuneaton in the afternoon, he would proceed the same night by road to Tamworth; and the next morning he would be out on the road, so soon, that he had the reputation, among his staff, of being the first man on the works. He used to proceed over the works from Tamworth to Stafford, walking the greater part of the distance; and he would frequently proceed that same evening to Lancaster, in order to inspect the works then in progress under the contract which he had for the execution of the railway from Lancaster to Carlisle. The journey which I have described from Havre, viâ London to Rugby, thence over the Trent to Stafford, and by railway to Lancaster, to inspect the Lancaster and Carlisle line, was a route which he very commonly followed."

[1] Mr. Murton's evidence.

CHAPTER VIII.

THE GREAT NORTHERN RAILWAY.

(A.D. 1847.)

Mr. Ballard.—A quaking bog.—Bridges in the Fen district.—Number of men employed.—The Brassey Shield.

IN this Chapter will be described one of Mr. Brassey's most important enterprises—namely, the formation of the Great Northern Railway. Some of the greatest difficulties depending upon the construction of this railway were solved by the intelligence of Mr. Stephen Ballard. But never were Mr. Brassey's qualities for choosing fit men, for appreciating their work as it proceeded, and for dealing wisely and kindly with them, more conspicuous than on this occasion.

It is noticed elsewhere how many of Mr. Brassey's chief agents rose from small beginnings, and became persons distinguished for their capability in the management of the greatest affairs. Mr. Ballard was no exception to this rule. He was intended for a nurseryman; was employed for two years in the gardens of Messrs. Lee and Kennedy of Hammersmith; and afterwards went to Hewell Grange to learn kitchen gardening. When he came to years of discretion, he determined to learn building, and spent three years as an apprentice to Mr. Lucy of Cheltenham. He was then appointed by Mr. Biddulph, the grandfather of Mr. Michael Biddulph, M.P., as manager of the Hereford and Gloucester Canal; and, after the completion of that work, he was made the resident engineer of the Middle Level Drain, part of the Great Bedford Level in the Fens. That work was on the point of completion when Mr. Ballard first became acquainted with Mr. Brassey.

The introduction was quite accidental, and took place at the railway station at Cambridge. Mr. Brassey was then returning to London from a visit of inspection to the country through which the Great Northern Railway was intended to pass. Mr. Brassey was not a man to be easily daunted by the difficulties of any undertaking; but anyone, who knows the Fen country, must be well aware that there were difficulties which would cause any contractor to feel very anxious about the prospects of an undertaking in a district of such a peculiar nature. The person who had introduced Mr. Ballard to Mr. Brassey was Mr. Smith, the agent of the contractor for the great Sluice of the Middle Level Bank. After these gentlemen had got into the railway carriage, Mr. Brassey mentioned that he had been looking over the line of the proposed Great Northern Railway, which was to pass through the fens; and with regard to a certain quaking bog which he had examined, he said. "You can stand upon it, and shake an acre of it together." He added, that he hardly knew what to do with the railway there, but that it *must* be carried over. He supposed that he must make a float, and float it over. Mr. Smith replied, "You had better get Mr. Ballard's advice as to the best way of passing over that quaking bog: he has had considerable experience in fen-work in our Middle Level Drain." The result of this interview was, that Mr. Ballard became the principal Agent for Mr. Brassey as regards the works that had to be carried out in the Fen country. The Fens which Mr. Ballard had to master were those adjacent to Wittlesea Mere, which is now drained and has good crops growing on what was once the bottom of it. The depth of the bog was no less than twenty-two feet, and it extended on a level for about three miles.

Mr. Ballard divided the Fen country which he had to deal with into three districts; and made a report to Mr. Brassey showing the different modes of treatment that should be applied to each district. Mr. Brassey immediately adopted the conclusions arrived at in that report. There were certain portable engines held to be necessary for the work—engines of a kind that were then very rarely used. One, however, had already been made by the well-known firm of Clayton and Shuttleworth, which engine, by the way, it had taken them six weeks to make, while now[1] they are turning out about four of these engines every day.

[1] A.D. 1872.

This engine had been sold to a farmer, but Mr. Ballard persuaded him to give it up upon payment of a certain sum. The next thing was to provide a large quantity of faggot-wood. For this purpose he caused 100 acres of faggot-wood to be cut down. A platform was then constructed of the stakes thus cut, placing them end to end the reverse way. Then, upon the first layer of stakes, a layer of peat sods was placed; afterwards stakes were placed longitudinally, upon which another layer of sods was put. Then a transverse layer of stakes; upon this a third layer of sods. By this time a sort of raft had been formed, upon which the soil was gradually piled up, giving the water time to run out. I will here avail myself of Mr. Ballard's own words:—

"The effect was to displace the water, but to leave the solid parts behind. This is not to be done by suddenly adding on great weight, but by a gradual and slow increase of weight, so that the water has time to escape without carrying away with it the solid matter. That plan has succeeded, and you may go to any height you like if you treat peat in that way."

A further difficulty, of much apparent magnitude, occurred in reference to the bridges which had to be constructed in the Fen country. Here again I had better give the constructor's own words:—

"I will explain the way in which we made the bridges— they were intended to be piled. The peat was twenty-two feet deep, and I pointed out to the engineer the difficulty of sufficiently bracing the piles, their tops being only about three feet above the soft bog.

"The piling was given up, and we made rafts of timber on which brick walls were built. These gradually sank, care being taken to so dispose the weight as to keep the walls perpendicular, and finally these walls were tested with rails of a greater weight than that of any train that could pass over them. We did not load too quickly, but left it; we put a little load on, and left it; and then the water had a chance of escaping. We only compressed the peat beneath the raft, without displacing it, for if we had once displaced it we must have gone down to the solid."

The result of this intelligence shown by Mr. Ballard in

conquering the difficulties of the Fen country, was, that what had been considered beforehand the most difficult part of the work, turned out to be the part most easily accomplished.

At that period certain difficulties arose as regards the financial part of the undertaking, and there was even some doubt whether a part of the works on the line should not be stopped. It was a notable feature in all Mr. Brassey's undertakings, that he could not bear to leave anything undone or half done. He might have made large claims for any delay occasioned by these financial difficulties. On the contrary, however, he arranged to facilitate the enterprise by taking the Company's mortgage Bonds in lieu of cash, as also did Mr. Ballard, who had engaged with Mr. Brassey to take a share of the profits—a percentage for his services. It may be noted that the Great Northern Company behaved with a very proper liberality to Mr. Ballard. He had sold his Bonds at a loss, and eventually they made that loss good to him.

The evidence of Mr. Ballard is as interesting and valuable as regards his superior, as it is respecting the men who worked under him.[1] It would be a pity to condense this or to give it in any other but his own words, as it affords such a good exemplification of character:—

It was impossible to walk along the line without receiving from him very valuable instruction—his experience was so very great, and he had such a comprehensive way always of looking at a difficult point. That was very remarkable and peculiar in him. If he came down to look at a line of Railway, he would walk over it, look at the crops of the country, and regard easy works as beneath his notice : he never looked at them ; but if there was a difficult point, as he could see by the section, then there was something to look at, and he would go and always put his thumb on the sore place : he would not waste his time in looking at little light matters that he knew were easy enough, but he always went straight to the sticking-place. He economized his time and brought his experience and judgment to bear where they were useful. He applied to engineering, that peculiar quality of concentration which is equally necessary in all other walks of life, in order to achieve success. He had great industry and always applied it in the right way. He would not fritter away his time upon trifles, when there was a more important thing to occupy his attention. He was very quick indeed in discovering beforehand, to the astonishment of those who had the management of the work, where there would be a pinch, and he would say, " If you do not do this, you

---

[1] Chapter V. p. 41.

will be wrong." He could look at a thing and foresee the state it would be in six months hence. His discernment was very acute, and he would point out to us where we were to provide for difficulties, which we might ourselves have overlooked; but regarding which the moment he mentioned them there could be no difference of opinion.

The largeness of these undertakings may be appreciated when, as in the present instance, it is found that at least from 5,000 to 6,000 men were employed. A slight mistake in the direction of the labour of these men, for even a few days, would be a very serious matter in a pecuniary point of view. This may illustrate to us the great responsibility which attached to the work of all those persons who had the guidance and governance of these great undertakings: for it is a loss to the State as well as to the individual employers of labour, when labour is in any way misplaced or misdirected.

Those who were interested in the Great Northern Railway and Mr. Brassey's other contracts acknowledged his services in a very splendid and artistic way. They collected nearly £2,000, and employed the proceeds in having full-length portraits of himself and Mrs. Brassey painted by the late Mr. Frederick Newenham. They also presented him with a silver-gilt shield, a very exquisite work of art, which was exhibited in the Exhibition of 1851. This shield was designed by Mr. H. P. Burt, and has in the centre the Brassey arms, surrounded by portraits, enamelled in gold, of twelve of the engineers under whose direction Mr. Brassey had executed important works. There are also twelve views of the principal works he had executed up to that period: and outside them a blue ribbon in enamel, bearing the names of thirty-six of Mr. Brassey's agents. The shield measures a yard in diameter, and was presented to him in March, 1851.

CHAPTER IX.

FINANCIAL MANAGEMENT.

System of accounts.—Mr. Tapp's opinion.—Monetary difficulties in Spain.—Paper money.—The Bilboa Bank.—Mr. Brassey's credit.—A Carlist chief.—Remuneration of agents.—Cheerfulness over financial losses.

MR. BRASSEY'S financial management of such great undertakings as those which have already been described, is very interesting and instructive from its exceeding simplicity. Mr. Tapp, who was Mr. Brassey's financial secretary and confidential adviser upon all monetary matters, has furnished conclusive evidence upon this point.

It appears that Mr. Brassey's mode of keeping the accounts of all his railway undertakings was purely local in its nature. On each contract the agent was responsible to Mr. Brassey for the money he received; and Mr. Brassey always looked to that agent to give him information, in London, as to the way in which the contract was being carried out. "He kept no regular check upon it, but simply noted that so much money had been sent to such and such a work. Beyond that no one knew anything of the account; he relied upon the cashier to keep the accounts, and he was supposed to audit them every month, and always to be in a position to give Mr. Brassey any information he required."

In contrasting this mode of keeping accounts with that adopted by any of the great spending departments of Government, such as the Admiralty, it is to be noticed that they have not only a local account of the same nature as that which was furnished to Mr. Brassey, but also a general account for the whole service, minutely kept in London. Such a system may be necessary for a great Government department; but it is one which it would have been very unadvisable for Mr. Brassey to adopt. His secretary, however, admits that "it requires a man like Mr. Brassey to carry out such a system as he adopted, because it is one particularly liable to abuse. He was very fortunate in that respect, for I am not aware that any one of his representatives ever deceived him, or robbed him. But still, other people, who are not possessed of the same discernment of character as Mr. Brassey, might, in my opinion, be very much imposed upon by relying implicitly upon one person."

When asked whether the honest service which Mr. Brassey thus received from his agents was, in the witness's opinion, a proof of the general honesty of mankind, he thus replied:—"Not exactly that. I think it rather more shows that he placed so much confidence in those whom he employed, and put them as it were so much upon their honour, that they would not deceive him, and that people who might not have acted uprightly with other people, did so with him because they felt responsible to him, and also a certain amount of pride in being confided in by him to such an extent, that they really carried on the business as if it was their own."

I cannot but think that the system of trust which Mr. Brassey adopted uniformly, with respect to all those who worked under him, was such as would be generally successful if carried out with that perfect faith and completeness which he always manifested in these transactions.

There were few, if any, of the great undertakings in which Mr. Brassey embarked that gave so much trouble in respect of the financial arrangements as the Spanish railway from Bilbao to Tudela. The Secretary thus recounts the difficulties which they had to encounter:—

The great difficulty in Spain was in getting money to pay the men for doing the work—a very great difficulty. The Bank was not in the habit of having large cheques drawn upon it to pay money; for nearly all the merchants kept their cash in safes in their offices, and it was a very debased kind of money, coins composed of half copper and half silver, and very much defaced. You had to take a good many of them on faith. I had to send down fifteen days before the pay day came round, to commence getting the money from the Bank, obtaining perhaps £2,000 or £3,000 a day. It was brought to the office, recounted and put into my safe. In that way I accumulated a ton or a ton and a half of money, every month during our busy season. When pay week came, I used to send a carriage or a large coach, drawn by four or six mules, with a couple of civil guards, one on each side, together with one of the clerks from the office, a man to drive, and another a sort of stable man, who went to help them out of their difficulty in case the mules gave any trouble up the hilly country. It was quite an operation to get this money out. I was at the office at six o'clock, and I was always in a state of anxiety until I knew that the money had arrived safely at the end of the journey. More than once the conveyance broke down in the mountains. On one occasion the axle of our carriage broke in half from the weight of the money, and I had to send off two omnibuses to relieve them : I had the load divided, and sent one to one section of the line, and one to the other.

Q. Was any attempt made to rob the carriage?

A. Never : we always sent a clerk armed with a revolver as the principal guard. We heard once of a conspiracy to rob us; but, to avoid that, we went by another road. We were told that some men had been seen loitering about the mountain the night before.

The payment of wages to the men employed on this Spanish railway affords a good illustration of the service that was done, in an indirect manner, to many countries where British enterprise was introducing the railway system. On this occasion the Basques were taught the use of paper money. Mr. Brassey's secretary, whose difficulties in procuring and distributing a metallic currency have already been seen from his own description, naturally endeavoured to make use of the paper money as more portable and less difficult to count. This was done by using the notes of the Bank of Bilbao for 100 *reals*,[1] equal in value to about £1. It may be noticed, by the way, that the Bank increased its profits to such an extent, that the shares were doubled in price while Mr. Brassey was working in that country.

---

[1] Mr. Tapp remarks, "That the 100-*real* gold piece is worth exactly 250 pence, or 1,000 farthings; and if ever a decimal currency should be adopted in England, it would be the most convenient form for the unit of value, as it would not disturb the relative values of the lower denomination of coins used by the poor, *a great difficulty in framing any new system of currency*.

"It would also represent exactly five dollars, and so pass without difficulty in the United States, South America, Portugal, and China, and to some extent in India and the Indian Archipelago. It would represent twelve and a half Austrian silver florins, and pass in France, Belgium, and Italy as twenty-six francs.

"I cannot understand how, in the many systems of uniform currency advocated by writers on finance, it has been lost sight of; the twenty-

FINANCIAL MANAGEMENT.

Inevitably, however, amid such a conservative people as the Basques, there was some difficulty at first in getting this paper currency into circulation.

When they came to receive an account, and I gave them so many hundred reals in notes, and the balance in silver; they would take up the silver, and stand waiting and say, "This is not the amount of my account." I said, "You have not taken up the notes." "What are these—where am I to get the money for them?" "Go to the Bank, and you will get the money." Then they went immediately to the Bank and changed the notes—the first time hesitatingly, but after the second and third times, finding that they were always paid, they took them home, and kept them till the next market-day.

During Mr. Brassey's occupation of that part of the country, for occupation it may be called, seeing that he and his partners had 10,000 men in their employment, he succeeded in obtaining from Madrid a supply of 100-*real* pieces, which are very difficult to obtain at Bilbao. There is one circumstance mentioned by this financial agent, which gives a good insight into the manner in which credit was attached to Mr. Brassey's name. By some unavoidable accident Mr. Tapp was once left at Chambéry, during the construction of another railway, without any arrangement having been made for paying the men; no authority had been given to him to sign cheques, and the person who could sign them was absent. He says:—

I went to the Bank, and told them how I was left, and that I had a large " Pay " to make, and asked them to let me have the money on my own cheque, and I was allowed to draw as much as £28,000 on one occasion.

The natural financial difficulties of constructing a railway in Spain were added to by the strange kind of people Mr. Brassey's agents were obliged to employ. One of the sub-contractors was a certain Carlist chief whom the Government dared not arrest on account of his great influence. Mr. Tapp thus relates the Carlist chief's mode of settling a financial dispute :—

five franc piece, which public opinion seems drifting towards, will prove most inconvenient for division into small sums in countries inhabited by the Anglo-Saxon race, by whom the greatest part of the trade of the world is carried on."

When he got into difficulties Mr. Small, the district agent, offered him the amount which was due to him according to his measured work. He had over 100 men to pay, and Mr. Small offered him the money that was coming to him, according to the measurement, but he would not have it, nor would he let the agent pay the men. He said he would have the money he demanded; and he brought all his men into the town of Orduna, and the men regularly bivouacked round Mr. Small's Office:—they slept in the streets, and stayed there all night, and would not let Mr. Small come out of the Office till he had paid them the money. He attempted to get on his horse to go out—his horses were kept in the house (that is the practice in the houses of Spain); but when he rode out, they pulled him off his horse and pushed him back, and said that he should not go until he had paid them the money. He passed the night in terror, with loaded pistols and guns, expecting that he and his family would be massacred every minute, but he contrived eventually to send his staff-holder to Bilbao on horseback. The man galloped all the way to Bilbao, a distance of twenty-five miles, and went to Mr. Bartlett in the middle of the night, and told him what had happened. Mr. Bartlett immediately got up and went to the military Governor of the town, who immediately sent a detachment up to the place to disperse the men. This Carlist threatened that if Mr. Small did not pay the money, he would kill every person in the house. When he was asked, "Would you kill a man for that?" he replied, "Yes, like a fly,"[1] and this coming from such a man who, as I was told, had already killed fourteen men with his own hand, was rather alarming.

It is not surprising to find that Mr. Brassey and his partners lost a very large amount upon the Bilbao Railway. This loss, however, must not altogether be attributed to the difficulty of dealing with the Spanish people. It must be confessed that the estimates, in this instance, were wrongly framed. It was supposed that the material to be dealt with was half earth, and half rock; but instead of that it turned out to be mostly hard rock. Then again the climate is a very wet one. Moreover, not only wet days, but fête days ought to have been allowed for. The consequence was, that Mr. Brassey's people were not able to work more than 200 days out of the 365 in the year.

The general results of this witness's evidence are peculiarly interesting on account of the light they throw upon Mr. Brassey's character. It was only after the secretary's return from Spain that Mr. Brassey gave him orders to make up a list of all his property. The secretary had to obtain his information from other people, such as the secretaries of the companies in which Mr. Brassey held shares.

---

[1] "Como una mosca."

"This took a long time to complete, so much so, that after I thought it was completed, in the next two years amounts kept cropping up, and I found, from some correspondence, that there was money that we knew nothing about, but we ultimately got it put into books in a regular way, and by degrees got the account correct and perfect."

This ignorance of his resources may be thought to indicate some carelessness on Mr. Brassey's part; but the truth is, that he was a man very indifferent to the possession of money. His mind was always occupied in getting the work through that he had undertaken, and there was a certain apparent carelessness about his own private affairs which only gives us a higher notion of the unselfishness of the man. It was not connected with any deficiency of financial ability. Mr. Brassey knew thoroughly well a good investment from a bad one; but he never seemed to take the trouble to think about investments. As his secretary observes, " I remember urging him very much to sell some shares when they were at a large premium, but he would not do it. He seemed to consider it a thing unworthy to be attended to, as if he thought some one else would lose by it, and that he would be taking the profit away from some one else; or that, having gone into the thing from its origin, and being to some extent responsible for its initiation, he ought to see it through, without getting out of it over some one else's shoulders. I do not know exactly why, but I never could get him to sell any of the shares which he subscribed for previous to the panic of 1866.

" I believe he felt that he had been one of the promoters, and if he got out of it, others might get out, and he would not abandon a ship in difficulties."

We learn from this witness the way in which Mr. Brassey remunerated his agents. "It was a system of paying sometimes by salaries and sometimes by a percentage on profits. The salaries which Mr. Brassey gave were decidedly not large; but he assigned to his principal agents a percentage upon the profits of the undertaking. In some instances these agents received cheques varying from £3,000 to £16,000. Indeed, several of those gentlemen who served under him succeeded in realizing fortunes."

This witness, like many others, speaks of the exceeding

cheerfulness of his employer, especially when he was tried by difficulties and disasters in the work, and by considerable pecuniary losses. He says:—

"I remember Mr. Bartlett, who had known Mr. Brassey as a younger man than I did, telling me that Mr. Brassey never appeared so happy as when he had lost £20,000. Whether it was that he made an effort at cheerfulness to throw it off his mind, I cannot say; but Mr. Bartlett said that he used to rub his hands, and that anyone would have supposed that he was delighted rather than otherwise. I remember, even at the time of the panic, when things were at the worst, Mr. Brassey saying one night, at the Westminster Palace Hotel, "Never mind, we must be content with a little less; that is all." That was when he supposed he had lost a million of money."

During the construction of the Bilbao line, shorly before the proposed opening, it set in to rain in such an exceptional manner that some of the works were destroyed. The agent telegraphed to Mr. Brassey to come immediately, as a certain bridge had been washed down. About three hours afterwards another telegram was sent, stating that a large bank was washed away; and, next morning, another, stating the rain continued, and more damage had been done. Mr. Brassey, turning to a friend, said, laughingly: "I think I had better wait until I hear that the rain has ceased, so that when I do go, I may see what is *left* of the works, and estimate all the disasters at once, and so save a second journey."

No doubt Mr. Brassey felt these great losses that occasionally came upon him much as other men do; but he had an excellent way of bearing them, and, like a great general, never, if possible, gave way to despondency in the presence of his officers.

This witness concludes his evidence in these words:—

"As to mere money-grubbing, he had not any of that in his composition, but he knew the value of money as well as anyone, and how far a pound would go; but he had no greediness to acquire wealth, and he was always willing to give away a portion of his profits to anyone who was instrumental in making them, and that to a remarkable extent."

CHAPTER X.

FINANCIAL DIFFICULTIES.

(A.D. 1866.)

Difficulties in 1866.—Victoria Docks.—Danish contracts.—Lemberg and Czernowitz Railway.—Evesham and Redditch Railway.—Warsaw and Terespol Railway.—Queensland Railway.—Great Western Branches.—Great Eastern Railway.—Barrow Docks and Runcorn Bridge.—Mr. Wagstaff.—Chevalier Ofenheim.—The Cross of the Iron Crown.

HITHERTO the narrative of Mr. Brassey's labours has chiefly been one of unexampled success, and there has been very little in the way of adversity to vary the narrative. The falling of a viaduct—earthwork turning out to be of a much more difficult kind than was expected—a scarcity of labour, and the like, were only momentary evils, scarcely sufficient to chequer the continuous success.

But in the year 1866, Mr. Brassey had to encounter an amount of financial difficulty and trouble which was sufficient to overwhelm almost any man, and which, though he bore the weight of it with great fortitude, had, in the opinion of some of those who knew him best, a considerable effect upon his health and life.

When we come to know the secret history of any great firm, or of any man whose financial enterprises have been very large and extensive, we almost always find that there has been a period of great difficulty and great peril—in short, a financial crisis. It is not altogether unpleasing to men of smaller means, when contemplating their own difficulties, to find that their great compeers have had to struggle through similar difficulties, and to overcome similar dangers to those which they themselves have had to encounter. The private crisis, too, generally comes at a time of public crisis—perhaps of panic—and when ordinary financial resources are for the moment unavailable.

Difficulties have an ingenious way of coming together, as if on set purpose, and by pre-arrangement amongst themselves, at the most inopportune conjuncture. We see this in the lives of great statesmen, and great generals, as well as in the lives of those who take the lead in commercial enterprise. And in the year 1866 the difficulties which Mr. Brassey had to encounter, advanced in a compact body against him. I will recount them one by one.

In the first place, there were liabilties in connection with the Victoria Docks, to the amount of £600,000.

Then there were the Danish contracts. In these contracts for certain railways in Denmark Mr. Brassey was associated with Messrs. Peto and Betts. The partners had obtained a loan from the General Credit Company of £300,000, upon an agreement to pay it back at the end of three years. This sum became due in 1866, at the time of the failure of Sir Morton Peto and Mr. Betts. The firm of Messrs. Peto, Brassey and Betts had also at that time large engagements outstanding for rails for the Danish contracts, involving very heavy liabilities. In fine, their liability for the Danish works may be stated at about £800,000.

Then there was the Lemberg and Czernowitz line. Mr. Brassey had received bonds from the Company to the amount of £1,200,000; but, at the moment, these bonds were worth very little more than so much waste paper. An effort was made to place them in a foreign market, which succeeded only to the extent of £13,000; and from that time they were perfectly unsaleable. Meanwhile Mr. Brassey had to pay from £40,000 to £50,000 a month for wages alone on that line.

There was also the Evesham and Redditch Railway for which Mr. Brassey was entirely paid in shares; and I must also mention the Warsaw and Terespol Railway, the payment for which was to a great extent made in bonds of which very few could be sold before the line was opened.

The Queensland Railway also involved heavy liabilities, and until the settlement of the account by the government of the colony some time later, nothing could be counted upon from this source.

Then there were several contracts in the West of

England, for lines in connection with the Great Western Company, such as the Wellington and Drayton line, the Nantwich and Market Drayton line, and many others. For these works Mr. Brassey was paid in the shares of the Company, which were at that juncture totally unsaleable. I may also mention that the Great Eastern Railway Stock, largely possessed by Mr. Brassey, was entirely useless as a financial resource.

The list of difficulties is not yet complete: there was also a heavy loss going on at the Barrow Docks and at Runcorn Bridge, amounting to £44,000.

In fact, the liabilities coming upon Mr. Brassey in that eventful year were so heavy that his property, of every kind whatsoever, was "largely committed." Such a man as Mr. Brassey was sure to have made devoted friends; and they were not wanting to him on this occasion. I must especially mention Mr. Wagstaff, who was of eminent service to him. He had been for many years a most intimate friend of Mr. Brassey's, whose confidence in his judgment, and reliance upon his friendship, were such that Mr. Brassey could not bear to undertake anything, or, at any rate, to prosecute any undertaking, without immediately informing Mr. Wagstaff, and seeking for his aid and counsel. It is needless to give an account of all the details of those transactions by which Mr. Brassey was enabled to push his way through his great difficulties during that critical period. One remarkable circumstance, however, deserves special notice—namely, that in spite of these financial difficulties, be persevered throughout that year in his old system of bringing works rapidly to a conclusion. Mr. Tapp says, "That Mr. Brassey was recommended by Mr. Glyn, Mr. Wagstaff, and indeed by almost all his friends, to delay the Lemberg and Czernowitz works." The reader will remember that these required from £40,000 to £50,000 a month for wages. "Still Mr. Brassey would go on. He would not stop the work; and it was a fortunate thing that he carried them on, because he was paying the interest of the shareholders, which amounted to over £120,000 a year. He had to pay them until the line was opened, when the Government guarantee came into force; and instead of being finished in January, it was finished in the previous September or October—four months before the contract time, and that added very much to his prestige in Austria." It not only added to his prestige, but it brought into play large funds which had hitherto been unavailable, for the Anglo-Austrian Bank now found that they could do a profitable business by selling the bonds, of which Mr. Brassey possessed more than a million. Certainly Mr. Brassey's bold, we may almost say audacious, perseverance in his accustomed course of finishing work as quickly as possible, at any loss, and at almost any hazard, was amply successful on this critical occasion.

The difficulties under which this Austrian line was completed were very great. These were not merely financial, but such as must arise from a state of war.

Mr. Brassey was admirably seconded in his efforts to complete this railway by Mr. Victor Ofenheim,[1] director-general of the Company, who also acted as one of Mr. Brassey's advisers on Austrian questions. The works were at that time progressing chiefly at Lemberg, five hundred miles from Vienna. The difficulty was how to convey the money from Vienna to Lemberg to pay the men. The intervening country was occupied by the Austrian and Prussian armies, who were on each side of the line, that is on that part between Cracow and Lemberg; for Mr. Ofenheim had succeeded without much difficulty in getting the money carried on the Northern Carl-Ludwig Railway as far as Cracow. However he was full of energy, and was determined to get on somehow or other. They said that there was no engine; that they had all been taken off; but he went and found an old engine in a shed. Next he wanted an engine-driver, and he found one, but the man said he would not go, for he had a wife and children; but Mr. Ofenheim said, " If you will come, I will give you so many hundred florins, and if you get killed I will provide for your wife and family." They jumped on to the old engine and got up the steam. They then started and went at the rate of forty or fifty miles an hour, and passing between the sentinels of the opposing armies; and Mr.

---

[1] Now the Chevalier d'Ofenheim.

Ofenheim states that they were so surprised that they had not time to shoot him. His only fear was that there might be a rail up somewhere. But he got to Lemberg, and that was the saving point of the line—they made the 'pay'—otherwise the men would have gone away to their homes, and the line would have been left unfinished through the winter, and they would have had to wait until the next spring before they could have returned again, but that difficulty being overcome got the line duly opened. Mr. Ofenheim's conduct on this occasion is a notable instance of the influence Mr. Brassey exercised over those who worked with him, as well as those who worked for him: for Mr. Ofenheim had become a devoted friend, as well as a skilful and daring representative of Mr. Brassey. The Emperor of Austria, with that appreciation shown by monarchs for devoted service—a thing they naturally very much approve of—was much struck by what he had heard of this daring feat in getting to Lemberg, and sent for Mr. Ofenheim, and asked this pertinent question: "Who is this Mr. Brassey, this English contractor, for whom men are to be found who work with such zeal, and risk their lives?" The answer must have been satisfactory, for the Emperor said Mr. Brassey must be a very powerful man, and sent him the Cross of the Iron Crown.

CHAPTER XI.

MR. BRASSEY'S WEALTH.

How Mr. Brassey employed his capital.—An instance of liberality.—Temporary embarrassments.—Causes of wealth.

AFTER giving an account of the financial difficulties which Mr. Brassey had to encounter at a late period of his career, it will be well to say a few words about the financial result, which, as the world knows, was a very successful one.

The acquisition of a great fortune by any man is not a thing which is intrinsically pleasing to the rest of mankind. Accordingly, they have invented divers sayings against the acquirers of large fortunes; such as that Arabic one, "Happy are the sons of those fathers who do not go to a good place," meaning thereby that great riches are seldom inherited from those who have gained them with entirely clean hands.

The more just observers of mankind have been wont to say, that their fellow-men are seldom more innocuously employed than in amassing wealth.

The truth is, that whatever fortunes Mr. Brassey and others of his calling accumulate are seldom or never ascertained and realized until their death, or until misfortune overtakes them in their lifetime. The capital that Mr. Brassey dealt with was never idle. As soon as any part of it ceased to be wanted for one great work, it was required for another, which either had to be commenced or was entering into a phase of full activity. It must be recollected, that a main object with Mr. Brassey was to furnish sufficient work for all that staff of skilled agents and for those large bodies of workmen whom he had collected around him. He had never the feeling of being a man of realized fortune—a millionaire, as we term it.

Of the numerous acts of Mr. Brassey's kindness and generosity, one may be quoted as an instance, viz. that of one of his old agents, of much merit and worth, who had unfortunately lost the whole of a competent independence, which he had acquired in Mr. Brassey's service. Mr. Brassey, anxious to give him the opportunity of recovering himself, confided to him several missions connected with new projects. In the last of these the agent was taken suddenly ill before reaching his destination, and died immediately after his arrival; almost at the very same time, his wife, whom he had left in good health in England, succumbed to a still more sudden attack. A family of six children were thus left orphans, and without any sort of provision. Mr. Brassey had already made an advance of several thousand pounds, for which he held as security a policy of insurance on the life of his agent. This he immediately relinquished in favour of the children, and further headed, with a substantial sum, a list of subscriptions made by the friends of the deceased for the orphans.

Though very liberal in relieving all cases of distress, which came within his immediate cognizance,[1] and especially those which originated amongst persons of his own staff, his name does not figure largely in subscriptions to public charities. He had other uses for his money, and these other uses were continually pressing upon him, but some notion may be formed of Mr. Brassey's habitual liberality, when I state that it is estimated that during his lifetime he gave away about £200,000. Occasionally, in the course of his life, as happens to most men of very extended affairs, he was greatly embarrassed, for the moment, to provide small sums, as we must call them—twenty or thirty thousand pounds, for instance—which were suddenly and urgently wanted in his business.

There were times at which, if he had died, he would have been found, comparatively speaking, a poor man. A man so situated can scarcely feel himself to be the possessor of millions, even though he might know that if his property were favourably realized at some particular time, it would amount to millions.

---

[1] See Letter 4, in Appendix.

The acquisition, however, of such a fortune as Mr. Brassey left behind him requires some explanation, though, as will be seen, it needs no apology.

There are two causes to be given which led to the accumulation of the wealth that Mr. Brassey left behind him. One was the small extent of his personal expenses. He was a man who hated all show, luxury, and ostentation. He kept up but a moderate establishment, which the increase of his means never induced him to extend.

The second and far more important cause was the immense extent of his business. That extent was gained not only by his intellectual qualities, but by his moral qualifications. Other men were very desirous of dealing with a man who was not only of known skill in his work, but who was of good repute for uprightness, for promptitude, and for going through thoroughly with anything which he had once begun. He never haggled or disputed, or sought by delay to weary people into his terms. His transactions were frank, distinct, and rapid; and there was no man who could less abide any loss of time in the completion of any of his enterprises. The success of such a parson is almost inevitable. As one of his enthusiastic admirers, who had been employed by him from the first, and knew him well, was wont to say, "If he'd been a parson, he'd have been a bishop; if a prize-fighter, he would have had the belt."

It was not from excessive gains in any one transaction, or even in several transactions, that his fortune sprang. It will, perhaps, surprise the reader to learn the small percentage of profit which accrued to him from all his enterprises, taken as a whole. It was, as nearly as possible, three per cent. He laid out seventy-eight millions of other people's money, and upon that outlay retained about two millions and a-half. The rest of his fortune consisted of accumulations.

CHAPTER XII.

MR. BRASSEY'S CONTRACTS.

(A.D. 1834-1870.)

APPENDICES are, I am afraid, in these days of multitudinous books and easy reading, very frequently skipped even by those persons who may be considered diligent readers, anxious thoroughly to understand everything about which they read. I have, therefore, resolved to put the following Table of Mr. Brassey's Contracts into the body of this work, rather than relegate it to the doubtful region of appendices.

It shows, more forcibly than any words of mine can show, how vast and various were the labours of Mr. Brassey. Indeed, I must own, that even after I had endeavoured to follow closely the records of his most laborious life, I was astonished at perceiving, in this condensed tabulated form, how great his labours had been.

The reader will also note how his work gradually increases upon him, and will now better understand how Mr. Brassey's attention was, by degrees, diverted from the actual superintendence of one or two works, to the general supervision of many great works going on at the same time.

There were periods in his career during which he and his partners were giving employment to 80,000 persons, upon works requiring seventeen millions of capital for their completion.

I subjoin a list of Mr. Brassey's contracts; for the compilation of which I have to thank my friend, Mr. Arthur Ricketts. So numerous were the contracts, that he is still uncertain whether there may not be some omissions hereafter to be supplied.

TABLE OF RAILWAY AND OTHER CONTRACTS

Completed by Mr. Brassey between the years 1834 and 1870; together with the names of the Engineers under whose direction the several works were executed; also the partners, if any, in each enterprise; the Agents who superintended the works, and the mileage of each contract.

Year.	Contract.	Partner.	Engineer	Agent.	Mileage.
1834	Branborough Road				4
1835	Grand Junction Railway		Mr. Locke, M.P., F.R.S.		10
1837	London and Southampton Railway (Branches and Maintenance)		{ Mr. Locke, M.P., F.R.S. Mr. Neuman }	Mr. Ogilvie and others.	36
1839	Chester and Crewe Railway		{ Mr. R. Stephenson, M.P., F.R.S. }	Mr. G. Meakin	11
	Glasgow, Paisley, and Greenock Railway		Mr. Locke, M.P., F.R.S.	Mr. Strapp and others	7
	Sheffield and Manchester Railway		Mr. Locke, M.P., F.R.S.	Mr. Dent and others	19
1841	Paris and Rouen Railway	Mr. Mackenzie	{ Mr. Locke, M.P., F.R.S. Mr. Neuman }	{ Mr. E. Mackenzie Mr. J. Jones Mr. Goodfellow Mr. Day and others }	82
1842	Orleans and Bordeaux Railway	{ Messrs. W. & E. Mackenzie }	M. Pepin Lehalleur	Mr. E. Mackenzie	294
1843	Rouen and Havre Railway	Mr. Mackenzie	{ Mr. Locke, M.P., F.R.S. Mr. Neuman }	{ Mr. Day Mr. G. Smith Mr. J. Jones Mr. Swanson Mr. Goodfellow }	8

RAILWAY AND OTHER CONTRACTS.

Year	Railway		Engineer	Contractor	Miles
1844	Lancashire and Carlisle Railway	{ Mr. Mackenzie { Mr. Stephenson	Mr. Locke, M.P., F.R.S.	Mr. G. Mould	70
	Colchester and Ipswich Railway	Mr. Ogilvie	{ Mr. Locke, M.P., F.R.S. { Mr. Bruff	Mr. Ogilvie	16
	Amiens and Boulogne Railway	{ Mr. W. Mackenzie. { Mr. E. Mackenzie.	M. Bazaine. Sir W. Cubitt, F.R.S.	Mr. E. Mackenzie	53
1845	Trent Valley Railway	{ Mr. Mackenzie { Mr. Stephenson	{ Mr. R. Stephenson, M.P., { Mr. Bidder . [F.R.S. { Mr. Gooch .	{ Mr. J. Jones { Mr. S. Horn	50
	Chester and Holyhead Railway	{ Mr. Mackenzie { Mr. Stephenson	{ Mr. R. Stephenson, M.P., { Mr. Ross . [F.R.S. { Mr. F. Forster	Mr. Woodhouse	81
	Ipswich and Bury Railway	Mr. Ogilvie	{ Mr. Locke, M.P., F.R.S. { Mr. Bruff	Mr. Ogilvie	24½
	Kendal and Windermere Railway	{ Mr. Mackenzie { Mr. Stephenson		Mr. G. Mould	12
	North Wales Mineral Extension Railway.	{ Mr. Mackenzie { Mr. Stephenson	Mr. Robertson	Mr. Meakin	5
	Caledonian Railway (1st Contract)	{ Mr. Mackenzie { Mr. Stephenson	{ Mr. Locke, M.P., F.R.S. { Mr. Errington	{ Mr. G. Mould { Mr. Woodhouse	125
	Clydesdale Junction Railway	{ Mr. Mackenzie { Mr. Stephenson	{ Mr. Locke, M.P., F.R.S. { Mr. Errington	Mr. Woodhouse	15
	Greenock Harbour	{ Mr. Mackenzie { Mr. Stephenson	{ Mr. Locke, M.P., F.R.S. { Mr. Errington	{ Mr. Goodfellow { Mr. Barnard	—
	Scottish Midland Junction Railway.	{ Mr. Mackenzie { Mr. Stephenson	Mr. Errington	Mr. Falshaw	33
	Scottish Central Railway.	{ Mr. Mackenzie { Mr. Stephenson	{ Mr. Locke, M.P., F.R.S. { Mr. Errington	Mr. Falshaw	46
1846	{ Lancashire and Yorkshire Railway { (Maintenance) { Ormskirk Railway	Mr. Field	Mr. Hawkshaw, F.R.S.	Mr. Day	93
		Mr. Mackenzie	Mr. Meek	Mr. Greene	30
	Shrewsbury and Chester Railway	{ Mr. Stephenson { Mr. Mackenzie	Mr. Robertson	Mr. Meakin	25
	Mineral Line (Wales)	Mr. Stephenson	Mr. Robertson	Mr. Meakin	6½
1847	Buckinghamshire Railway		Mr. Dockray	{ Mr. S. Horn { Mr. Goodfellow { Mr. Day	47½
	{ Birkenhead and Chester Junction { Railway.		Mr. Rendel, F.R.S.		17½

Year.	Contract.	Partner.	Engineer.	Agent.	Mileage.
	Haughley and Norwich Railway	Mr. Ogilvie	Mr. Locke, M.P., F.R.S.	Mr. P. Ogilvie	33
	Great Northern Railway		Mr. J. Cubitt	Mr. Bartlett / Mr. Milroy / Mr. Ballard	75½
1847 cont.	North Staffordshire Railway		Mr. R. Stephenson, M.P., F.R.S. / Mr. Bidder	Mr. J. Jones	48
	Shrewsbury Extension Railway	Mr. Mackenzie / Mr. Stephenson	Mr. Robertson	Mr. Meakin	3
	Trent Valley Stations	Mr. Mackenzie / Mr. Stephenson	Mr. Bidder and others	Mr. Holme	—
	Blackwall Extension Railway	Mr. Ogilvie	Mr. Locke, M.P., F.R.S. / Mr. Stanton	Mr. Burt	1¾
	Richmond and Windsor Railway	Mr. Ogilvie	Mr. Locke, M.P., F.R.S.	Mr. Evans	16½
	Rouen and Dieppe Railway	Mr. Mackenzie	Mr. Neuman / Mr. Murton	Mr. Benyon / Mr. C. Smith	31
	Chester Station	Mr. Mackenzie / Mr. Stephenson	Mr. Robertson	Mr. S. Holme	—
	Oswestry Branch Railway			Mr. Meakin	2
1848	Loop Line (L. & S. W. R.)	Mr. Ogilvie / Mr. Mackenzie	Mr. Locke, M.P., F.R.S.	Mr. Evans	7
	Caledonian Railway (2nd Contract) (Stations, Maintenance, &c.)	Mr. Stephenson?	Mr. Locke, M.P., F.R.S.		—
	Glasgow and Barhead Railway		Mr. Locke, M.P., F.R.S. / Mr. Locke, M.P., F.R.S. / Mr. W. Locke	Mr. Strapp	11
	Barcelona and Mataro Railway	Mr. Mackenzie		Mr. Robson	18
1849	Royston and Hitchin Railway		Mr. Locke, M.P., F.R.S.	Mr. H. Harrison	13
	Shepwreth Extension Railway		Mr. Locke, M.P., F.R.S. / Mr. Rendel, F.R.S.	Mr. H. Harrison	5
	Birkenhead Docks		Mr. Abernethy	Mr. Dent	—
1850	North and South-Western Junction Railway	Mr. Ogilvie	Mr. G. Berkeley	Mr. Evans	4
	Prato and Pistoja Railway		Italian Government	Mr. T. Woodhouse	10

RAILWAY AND OTHER CONTRACTS.

Year	Railway	Contractor	Engineer	Assistant	Miles
1851	Shrewsbury and Hereford Railway, and Maintenance	Sir M. Peto, M.P., & Mr. Betts	Mr. Robertson	Mr. W. Field	51
	Norwegian Railway	Sir M. Peto, M.P., & Mr. Betts	Mr. Bidder	{ Mr. Merrit Mr. Earle	56
	Hereford, Ross, and Gloucester Railway	Sir M. Peto, M.P., & Mr. Betts	Mr. Brunel, F.R.S.	Mr. Watson	30
	London, Tilbury, and Southend Railway	Sir M. Peto, M.P., & Mr. Betts	Mr. Bidder	Mr. White	50
	Victoria Docks and Warehouses	Sir M. Peto, M.P., & Mr. Betts	Mr. Bidder		—
	Warrington and Stockport Railway		Mr. Lister	Mr. Holland	12
	North Devon Railway	Mr. Ogilvie	Mr. W. R. Neale	{ Mr. Goodfellow Mr. P. Ogilvie	47
1852	Mantes and Caen Railway		{ Mr. Locke, M.P., F.R.S. Mr. Neuman Mr. W. Locke	{ Mr. Evans Mr. J. Jones Mr. C. Jones Mr. J. Milroy	113
	Le Mans and Mezidon Railway		{ Mr. Locke, M.P., F.R.S. Mr. Bergeron	Mr. Woodhouse	84
	Lyons and Avignon Railway	Sir M. Peto, M.P., & Mr. Betts	{ M. Talabot M. Thirion M. Molard	{ Mr. G. Giles Mr. Murton	67
	Dutch Rhenish Railway		Mr. Locke, M.P., F.R.S.	Mr. Ballard	43
	Grand Trunk Railway	{ Sir M. Peto, M.P., Mr. Betts, & Sir W. Jackson	Mr. Ross	{ Mr. Reekie Mr. Hodges Mr. Rowan Mr. Tait Mr. P. Ogilvie	539
1853	Crystal Palace and West-End Railway	Sir M. Peto, M.P., & Mr. Betts	Mr. Bidder	Mr. Watson	5
	Sambre and Meuse Railway		M. Declerq	{ Mr. H. Harrison Mr. T. Woodhouse	28
	Turin and Novara Railway		Italian Government	Mr. Hancock	60
	Hauenstein Tunnel		M. Etzel	M. Benyon	1½
	Royal Danish Railway	Sir M. Peto, M.P., & Mr. Betts	Mr. Bidder Mr. G. R. Stephenson	Mr. McKeon	75
1854	Arpley Branch Railway		Mr. Lister	Mr. Goodfellow	1⅝/7½
	Woodford and Loughton Railway	Sir W. Jackson, Messrs. Felk&Jopling	Mr. Bidder	Mr. H. Harrison	
	Central Italian Railway		Italian Government	{ Mr. Fell Mr. Jopling	52

Year.	Contract.	Partner.	Engineer.	Agent.	Mileage.
1854 cont.	Turin and Susa Railway	Mr. C. Henfrey	Italian Government	Mr. C. Henfrey	54
	Bellegarde Tunnel	Messrs. Parent & Buddicom	M. Talabot	Mr. Goodfellow	2¼
1855	East Suffolk Railway	Sir M. Peto, M.P., & Mr. Betts	Mr. G. Berkley	Mr. Watson	63
	Inverness and Nairn Railway	Mr. Falshaw	Mr. Mitchell	Mr. Falshaw	16
	Portsmouth Direct Railway	Mr. Ogilvie	Mr. Locke, M.P., F.R.S.	Mr. Evans	33
	Caen and Cherbourg Railway		Mr. Locke, M.P., F.R.S.	Mr. Milroy	94
	Coghines Bridge		Mr. W. Locke	Mr. C. Jones	
			Italian Government	Mr. Dent	—
1856	Woodbridge Extension Railway	Mr. Ogilvie	Mr. Bruff	Mr. Boys	10
	Elizabeth-Linz Railway	Sir M. Peto, M.P., & Mr. Betts	M. C. Keissler	Mr. G. Giles	49
1857	Leicester and Hitchin Railway	Mr. Field	Messrs. Liddell & Gordon	Mr. Horn	62½
				Mr. Harrison	
	Leominster and Kington Railway	Sir M. Peto, M.P., & Mr. Betts	Mr. Wylie	Mr. Field	14
	Minories Warehouses		Mr. Tite, M.P.	Mr. Holland	—
1858	Leatherhead, Epsom, and Wimbledon Railway	Mr. Ogilvie	Mr. Locke, M.P., F.R.S.	Mr. Ogilvie	10
	Worcester and Hereford Railway	Mr. Ballard	Mr. A. C. Crosse	Mr. Ballard	26½
	Inverness and Aberdeen Junction Railway	Mr. Falshaw	Messrs. Liddell & Gordon	Mr. Falshaw	22
	Bilbao and Miranda Railway	Messrs. Wythes, Paxton & Bartlett	Mr. Vignoles, F.R.S.	Mr. Bartlett	66
	Eastern Bengal Railway	Messrs. Wythes, Paxton & Henfrey	Mr. Hawkshaw, F.R.S.	Mr. C. Henfrey	112
1859	Cannock Mineral Railway	Mr. Field	Mr. Addison	Mr. J. Stephenson	10
	Crewe and Shrewsbury Railway		Mr. Locke, M.P., F.R.S.	Mr. Day	32½
	Salisbury Station	Mr. Ogilvie	Mr. Errington	Mr. Carswell	2
			Mr. Tolmé	Mr. J. Walker	
	Denny Branches		Mr. Locke, M.P., F.R.S.	Mr. Falshaw	3
			Mr. Errington		

RAILWAY AND OTHER CONTRACTS.

Year	Railway/Project	Engineer	Client	Contractor	Miles
1859 cont.	Victor Emmanuel Railway	Sir W. Jackson, Mr. Henfrey	Mr. Neuman, Mr. Ranco	Mr. Bartlett, Mr. W. Strapp, Mr. Blake, Mr. Edwards	73
	Ivrea Railway	Mr. C. Henfrey	Italian Government	Mr. Dent, Mr. Dixon	19
	Great Northern, Great Eastern, Great Southern Railways (New South Wales)	Sir M. Peto, M.P., & Mr. Betts	Mr. Whitton	Mr. Wilcox, Mr. Rhodes	54
1860	Salisbury and Yeovil Railway	Mr. H. Harrison	Mr. Locke, M.P., F.R.S.	Mr. H. Harrison	40
	Woofferton and Tenbury Railway	Mr. Ogilvie	Mr. Wylie	Mr. Mackay	5
	Wenlock Railway	Mr. Field	Mr. Fowler	Mr. Seacome	4
	Port Patrick Railway	Mr. Field	Mr. Blyth	Mr. Falshaw	17
	Stokes Bay Pier and Branch Railway	Mr. Falshaw	Mr. Fulton	Mr. Evans	2
	Harleston and Beccles	Mr. Ogilvie	Mr. Bruff	Mr. Boys	13
	Dieppe Railway (laying second road)	Mr. Ogilvie	M. Julien	Mr. R. Goodfellow	—
	The Maremma, Leghorn, &c., Railway	Mr. Buddicom	M. Pini	Mr. Jopling, Mr. C. Jones	138
	Jutland Railway	Sir M. Peto, M.P., & Mr. Betts	Danish Government	Mr. Rowan	270
1861	Disley and Mayfield Railway	Mr. H. Harrison	Mr. Errington	Mr. Harrison	3¾
	Knighton Railway	Mr. Field	Mr. Robertson	Mr. Field	12
	Nuneaton and Hinckley Railway	Mr. Field	Mr. Addison	Mr. J. Stephenson	5
	Shrewsbury and Hereford Railway (widening)	Mr. Field	Mr. Wylie, Mr. Clark	Mr. J. Mackay	51
	West London Railway (Extension)	Mr. Ogilvie	Mr. W. Baker	Mr. Evans	9
	Ludlow Drainage	Mr. Field	Mr. Curley	Mr. Mackay	7
	Severn Valley Railway		Mr. Fowler	Mr. Field, Mr. Day, Mr. Dent, Mr. Dowell	42
	South Staffordshire Railway	Mr. Field, Mr. Ogilvie	Mr. McClean, M.P., F.R.S.	Mr. Day	4
	Metropolitan Mid Level Sewer	Mr. Harrison	Mr. Bazalgette, C.B.	Mr. Harrison	12
1862	Ringwood and Christchurch Railway	Mr. Ogilvie	Captain Moorsom	Mr. Evans	8
	Kingston Extension Railway	Mr. Ogilvie	Mr. Galbraith	Mr. Evans	4

Year	Contract.	Partner.	Engineer.	Agent.	Mileage.
	Cannock Chase Railway	Mr. Field	Mr. Addison	Mr. Cooper	3
	Coalbrookdale Railway	Mr. Field	Mr. Fowler	Mr. Dent	5
	Ashchurch and Evesham Railway	Mr. Ballard	Messrs. Liddell & Gordon	Mr. Ballard	11
	Nantwich and Market Drayton Railway		Mr. Gardiner	Mr. Gallaher	11
1862 cont.	South Leicester Railway	Mr. Field	Mr. Addison	Mr. Mackay	10
				Mr. J. Stephenson	
	Tenbury and Bewdley Railway	Mr. Field	Mr. Wylie	Mr. Mackay	12
			Mr. Clarke		
	Wenlock and Craven Arm Railway	Mr. Field	Mr. Fowler	Mr. Dent	14
				Mr. N. Mackay	
	Ludlow and Clee Hill Railway	Mr. Field	Mr. Wylie	Mr. Mackay	6
	Llangollen Railway		Mr. Clarke		
	Rio Janeiro Drainage	Mr. Ogilvie	Mr. Robertson, M.P.	Mr. Gallagher	6
		Mr. Wythes and others	M. Gotto	Mr. Houcox	—
	Mauritius Railway		Mr. Hawkshaw, F.R.S.	Mr. Longridge	64
	Epping and Ongar	Mr. Ogilvie	Mr. Sinclair	Mr. Harrison	13
	Barrow Docks	Mr. Harrison	Mr. McClean, M.P., F.R.S.	Mr. Dent	—
	Runcorn Branch Railway	Mr. Field	Mr. W. Baker	Mr. Evans	9
	Tendring Hundred Railway	Mr. Ogilvie	Mr. Bruff	Mr. Boys	3
	Worm Drainage	Mr. Field	Mr. Curley	Mr. Mackay	—
1863	Sudbury, Bury St. Edmunds, and Cambridge Railway	Mr. Ogilvie	Mr. Sinclair	Mr. Bell	48
				Mr. Boys	
	Meridionale Railway	M. Parent	M. Grattoni and others	Mr. Smalls	160
		Mr. Buddicomb		Mr. C. Jones	
	Queensland Railway	Sir M. Peto, M.P., & Mr. Betts	Mr. Fitzgibbon	Mr. Charles	78¼
	North Schleswig Railway	Sir M. Peto, M.P., & Mr. Betts	Mr. Rowan	Mr. Willcox	70
				Mr. Louth	
1864	Epping Railway	Mr. H. Harrison	Mr. Sinclair	Mr. Butler	12
		Mr. Ogilvie			
	Letton Drainage	Mr. Field	Mr. Curley	Mr. Mackay	3½

RAILWAY AND OTHER CONTRACTS.

Year	Work	Engineer	Contractor (Client)	Miles
1864 cont.	Dunmow Railway	{ Mr. Ogilvie Mr. H. Harrison }	Mr. Sinclair	Mr. Butler	18
	Corwen and Bala Railway		Mr. Robertson, M.P.	{ Mr. Field Mr. Reid }	14
	Wellington and Market Drayton Railway	Mr. Field	Mr. Wilson	Mr. Mackay	16
	Enniskillen and Bundoran Railway	Mr. Field	Mr. Hemans	{ Mr. Day Mr. Drennan }	36
	Central Argentine Railway	{ Mr. Wythes Mr. Wheelwright Mr. Ogilvie }	Mr. E. Woods	Mr. Wheelwright	247
	Lemberg Czernowitz Railway		{ Mr. McClean Mr. Stileman M. Ziffer M. de Herz }	Mr. Strapp	165
	Viersen-Venlo Railway	Mr. Murton	M. Lange	Mr. Murton	11
	Delhi Railway	{ Mr. Wythes Mr. Henfrey }	{ Mr. Bidder Mr. J. Harrison Mr. Bidder }	{ Mr. C. Henfrey Mr. Mareiller }	304
1865	Chertsey Extension Railway	Mr. Harrison	Mr. Galbraith	Mr. Harrison	3
	Dee Reclamation Works	{ Mr. Field Mr. Meakin }	Mr. Bateman, F.R.S.	Mr. Meaken	—
	Evesham and Redditch Railway	Mr. Ballard	Mr. Richards	Mr. Ballard	18
	East London Railway	{ Mr. Wythes Messrs. Lucas, Brothers }	Mr. Hawkshaw, F.R.S.	Mr. H. Harrison	2½
	Hull and Doncaster Railway	Mr. Field	Mr. Harrison	Mr. Stephenson	16
	Hereford Loop Railway	Mr. Field	Mr. Clarke	Mr. Mackay	2½
	Hooton and Parkgate Railway	Mr. Field	Mr. Johnson	Mr. Mackay	5
	London and Bedford Railway	Mr. Ballard	Mr. Liddell	Mr. Ballard	36½
	Llangollen and Corwen Railway		Mr. Robertson, M.P.	Mr. Field	10
	Nantwich and Market Drayton (Widening)	Mr. Field	Mr. Gardiner	{ Mr. Reid Mr. Mackay }	11
	Boos and Barracas Railway	{ Mr. Wythes Mr. Wheelwright Mr. Vignoles, F.R.S. }	Mr. Coghlan	Mr. Simpson	3
	Warsaw and Terespol Railway	Mr. Ogilvie	Russian Government	Mr. H. Vignoles	128

Year	Contract	Partner	Engineer	Agent	Mileage
1865 cont.	Chord Line (India)	Mr. Wythes { Mr. Perry	Mr. Rendel	Mr. Perry	147
	Calcutta Waterworks	Mr. Wythes { Mr. Aird	Mr. Purdon { Mr. Lewis	Mr. Paton	—
	Ebbw Vale Railway	Mr. Field	Mr. Gardiner	Mr. Mackay	2¾
	Thames Embankment	Mr. Ogilvie { Mr. Harrison	Mr. Bazalgette, C.B.	Mr. Harrison	—
	Kensington and Richmond Railway (and Spurs)	Mr. Ogilvie { Mr. Harrison	Mr. Galbraith { Mr. Tolmé	Mr. Evans	7
1866	Christchurch and Bournemouth Railway	Mr. Ogilvie	Mr. Strapp	Mr. Ogilvie	4
	Moreton Hampstead Railway	Mr. Ogilvie	Mr. Margary	Mr. Crossley { Mr. Field { Mr. Day	12
	Bala and Dolgelly Railway		Mr. Wilson	Mr. Drennan	18
	Sirhowy Railway	Mr. Field	Mr. Sayer	Mr. Mackay	2
	Wolverhampton and Walsall Railway	Mr. Harrison { Mr. Ogilvie	Mr. Addison	Mr. Harrison	7½
1867	Czernowitz Suczawa Railway		M. Ziffer { M. de Herz	Mr. Strapp	60
	Kronprinz Rudolfsbahn	M. Klein { M. Schwarz	M. F. Kagda	M. Fülsch	272
1868	Silverdale Railway		Mr. Forsyth	Mr. Field { Mr. Mackay	13½
1869	Nepean Bridge	Sir M. Peto, M.P., & Mr. Betts	Mr. Fowler	Mr. Willcox	—
	Callao Docks		M. Alléon	Mr. Hodges	—
1870	Vorarlbergbahn	M. Klein { M. Schwarz	M. W. Paravicini	Mr. Fölsch	55
	Suczawa and Jassy Railway		M. Ziffer	Mr. Strapp { Mr. Edwards	135

CHAPTER XIII.

THE ITALIAN RAILWAYS.

(A.D. 1850-53.)

Turin and Novara Railway.—Count Cavour and Mr. Giles.—Proposed Lukmanier Pass Railway.—Turin and Susa Railway.—The Victor Emmanuel Railway.—Bartlett's boring machine—Buffalora Extension Railway.

IT would far exceed our limits to enter into a detailed account of every great enterprise which Mr. Brassey undertook, either by himself, or in partnership with others. There are, however, certain remarkable points connected with several of these undertakings, which ought not to be passed over in silence.

I will begin with the railway from Turin to Novara. Count Cavour had intimated a strong desire to bring capital into his country for the purpose of constructing railways; and he naturally looked to Englishmen for assistance in attaining this desirable object. His wishes were responded to by our countrymen. Negotiations were accordingly commenced with the Count; and Mr. Brassey, Mr. Frank Mills, and Mr. Netlam Giles were the contractors who, under the sole name of Mr. Brassey, arranged the conditions of a concession from the Piedmontese Government for the line from Turin to Novara.

The preliminary agreement entered into with M. Paleocapa, the Minister of Public Works, was as follows:—That the Piedmontese Government was to subscribe a fourth of the capital, Mr. Brassey a fourth, the Provinces a fourth, and the public a fourth, no interest being guaranteed.

The Piedmontese public, however, did not understand railway matters, and were totally disinclined to subscribe for their share. Count Cavour sent for Mr. Giles one morning, and said, "We are in a difficulty: the public have subscribed for very few shares; but I am determined to carry out the line, and I want to know if Mr. Brassey will take half of the deficiency, if the Government will take the other half?" A promise was given on Mr. Brassey's part, that he should do so,—a noticeable instance again of how Mr. Brassey's partners as well as his agents could venture to act for him on very critical occasions. It was shortly afterwards announced that the subscriptions were covered; the Piedmontese then took heart, and applied for a very much larger amount of shares than that which had been originally offered to them. Count Cavour made an appeal to the promoters of the railway, saying "The public are now crying out that they cannot get a share, and the shares are at a good premium. Will you give up some shares, as I am anxious to whet their appetite for other enterprises by letting them taste a profit on their first speculation?' Cavour was asked, "How many do you want, supposing that the Government will give up as many?" He replied, "Will you give up 2,000?" He was assured that Mr. Brassey did not care about "jobbing the shares," and there was no doubt he would do what the Count wished. The matter was shortly afterwards discussed with Mr. Brassey, and he willingly gave up the 2,000 shares, they being then at more than £2 premium.

This railway from Turin to Novara was a very successful undertaking, as the traffic proved far more remunerative than even the original promoters had estimated, and the line was completed for a less sum of money than had been expected. Mr. Brassey, in discussing these transactions made this notable remark, "*That railway has been completed for about the same money as was spent in obtaining the Bill for the railway from London to York.*" The length of the Turin and Novara Railway was 60 miles.

The total charge of the Sardinian Government against Mr. Brassey for the concession was exactly £100.

It is therefore not wonderful that Mr. Brassey had the opinion that the system of concessions by a Government to Companies was very superior to the Parliamentary system which is adopted in this country.

Count Cavour said to Mr. Brassey, shortly after the

railway was in operation, "I am told the line *per se* is yielding 14 per cent.; and yet there was a time when I could not induce my Piedmontese to take a share!"

The successful construction of this line from Turin to Novara brought, as might be expected, an immense number of applications from all parts of Piedmont for the extension of the railway system. In fact, there was, as I am informed, "a perfect *furore* for the construction of railways throughout the country." Mr. Brassey was concerned in several of these enterprises;[1] but it is not necessary to enter into any details respecting their construction. Count Cavour was very grateful to Mr. Brassey, and repeatedly thanked him for the liberal and spirited way in which he had responded to the wishes of the Piedmontese Government.

The proceedings which took place in reference to the proposed formation of what was called the Lukmanier Line, from Locarno on the Lago Maggiore over the Lukmanier Pass to the Union-Suisse Railway at Coire, are worth noting; I will give them in the words of Mr. Giles:—

"It so happened that I had been interested in the original concessions from the Cantons Grisons and Tessin, of the Lukmanier Line, and at an interview between the Count and Mr. Brassey, Cavour said, 'Now, Mr. Brassey, I want you to do something for us in this matter, in which I take a great interest.' Mr. Brassey replied, 'Really, I know nothing about it; but,' pointing to me, 'Here is a man who knows everything about it—he has spent time and money upon it, promoting it. Perhaps your time, therefore, would be saved by explaining your views to him, and if I can assist them I shall be happy to do so.' So the Count said to me, 'Come and see me to-morrow morning.' I may mention that it was not unusual for Count Cavour to see people in the summer-time at five o'clock in the morning. My appointment was at six o'clock. I waited upon him as appointed. We then discussed the Lukmanier, and came to an arrangement. I said, 'There are no "surveys" in this matter, or no reliable surveys—they are all made by the people in the country. Will you share part of the

---

[1] See List of Railways, *ante*, pp. 84-92.

expense of a definitive survey?' He replied, 'I do not think, in the present position of matters, it can be done. It is in Switzerland; and the Swiss are so touchy about any interference of a foreign Government, that I think our doing so would have a prejudicial rather than a beneficial effect; but I should be glad if Mr. Brassey can see his way to making them without any assistance from us.' I spoke to Mr. Brassey about it, and the surveys were made in the spring of 1868, and Count Cavour was asked to meet Mr. Brassey and the promoters at Coire, for the purpose of inspecting the line.

"Cavour came to Coire on July 27, 1858. Mr. Brassey had fully intended to accompany the Count over the line, but unfortunately could not do so, as he had to complete the line to Cherbourg, which the Emperor Napoleon was to open on August 5. Immediately Cavour came to Coire, it set in to rain in torrents, and he was delayed a day. Mr. Giles waited upon the Count by appointment at six o'clock on the succeeding morning, and found him reading Macaulay's 'History of England.' He said, 'Wonderful and delightful book this; it is as exciting as a champagne breakfast.' They then discussed the proposed Lukmanier line, and the Count said, 'I very much regret Mr. Brassey is not here, as I have looked forward to the pleasure of going over the line with him, and thoroughly understanding how he proposes to construct the two sections, and the carriage road over the mountain. I am already acquainted, through M. Sommeiller, that Mr. Brassey thinks it better to make a good tunnel even in fifteen years than a bad one in six years. I think so too; indeed, I shall be disposed to accept whatever Mr. Brassey proposes, as I have full confidence in his opinion. I should like very much to go over the line with him; and if you will inform me when he will be at Coire, I will do my best to return, and accompany him over the line, as I am most anxious to have my lesson from the most experienced contractor in Europe, and so be able to discuss the question *au fond*, and with a full knowledge of the facts.'"[1]

There was a grand dinner at Coire in honour of Count

---

[1] Mr. Netlam Giles's evidence.

Cavour the same day, at which the Count said: "Mr. Brassey is one of the most remarkable men I know; clear-headed—cautious, yet very enterprising—and fulfilling his engagements faithfully. We never had a difficulty with him. He would make a splendid Minister of Public Works; and," he added, laughing, "if report be true, he understands the Finance Department equally as well."

This praise from Count Cavour was the praise of a man who was himself a consummate master in the management of affairs. Mr. Giles, who throughout these transactions was in frequent communication with the Count, says of him, "as a man of business, I never met his equal, except in Mr. Brassey."

After the revolution of 1848, political considerations, and probably strategical views, entered very largely as important circumstances in the arrangement of the system of Piedmontese railways.

It was not only Count Cavour, who was then the Finance Minister, who was anxious for the introduction of British capital for the purpose of railway construction. The Prime Minister, D'Azeglio, and the Minister for Public Works, M. Paleocapa, were also most anxious to promote the same kind of enterprise.

In 1850 a railway was planned to extend the Turin and Genoa line from Turin to Susa by Mr. Charles Henfrey. Previously there had been before the Piedmontese Government the project of tunnelling through Mont Cenis. Indeed, as far back as 1841, this project had been brought forward by Mons. Medail; and later, the Chevalier Maus had prepared a detailed project for a railway and tunnel, and had been allowed to make experiments with a boring machine of his own invention at the Government works. The condition of the national finances of Piedmont, however, entirely prevented the adoption of so great a scheme; and the Government, therefore, cordially received the proposal, made through Mr. Henfrey, of constructing a cheap line of railway between Turin and Susa. Its object was to facilitate the means of communication with France —Susa lying at the foot of the Mont Cenis Pass.

"By the construction of this line," as Mr. Henfrey ob-

H

serves, "railway communication would be complete from the Alps to the Mediterranean, and the first link in the chain of international communication with France and the West of Europe would be forged." A contract for its construction was made between the Piedmontese Government and Messrs. Brassey, Jackson, and Henfrey. The Piedmontese Government engaged to take one half the shares, and the contractors the other half. Moreover, the Government undertook the surveillance of the works, and to work and maintain the line at fifty per cent. on the gross receipts.

In continuance of the same great plan of making railway communication between Piedmont and France, a survey was made by Messrs. Brassey, Jackson, and Henfrey, on the northern side of the pass of Mont Cenis, down the valley of the Arc, to Chambéry, and thence to the French frontier at Culoz.

The Piedmontese Government had promised to give their support to this project. But here a French company stepped in, and proposed to Messrs. Brassey, Jackson, and Henfrey, that if they would withdraw their demand for a concession, they (the French company) would ensure to them the execution of the works, which offer was accepted. It was called the Victor Emmanuel Railway, in honour of the King of Sardinia.

The works on this line were commenced in 1855 and completed in 1858, under the superintendence of Mr. Thomas Bartlett. "It was," Mr. Henfrey says, "during the construction of this railway that Mr. Bartlett first brought into use his machine for boring rock, the principle of which was appropriated, with but faint recognition of his claims, by the Italian engineers, for the boring machinery of the Mont Cenis Tunnel. It may safely be said that, in the absence of such a machine for facilitating the boring of rock, the Sardinian Government would not have undertaken so gigantic a work as the Mont Cenis Tunnel—12,220 mètres, or $7\frac{1}{2}$ miles long, through solid stone; and it is gratifying to find that the importance of Mr. Bartlett's invention has been fully appreciated in France, as will be seen by the following extract from a brochure, entitled "Géologie des Alpes et du Tunnel des

Alpes," recently published by the eminent geologist, M. Elie de Beaumont:[1]—

"En 1855, un Anglais, M. Bartlett, construisit une machine perforatrice, qui fut essayée avec un plein succès à Gênes et à Chambéry.

"Au premier aspect on croyait avoir devant soi une simple locomotive; mais au piston de la machine à vapeur s'ajoutait un second piston plein d'air, dont la tige était armée d'une barre à mine. L'air faisait matelas, et empêchait les chocs trop brusques de se transmettre au piston moteur. La barre à mine frappait jusqu'à 300 coups à la minute.

"Le problème de la perforation mécanique était résolu; mais on ne pouvait raisonnablement songer à utiliser une machine à vapeur dans un trou d'une profondeur de plusieurs kilomètres. Le peu d'air respirable que l'on aurait pu envoyer aux ouvriers aurait été bien promptement vicié.

"Ici intervint l'idée féconde de l'emploi de l'air comprimé comme force motrice en remplacement de la vapeur. Rien n'était plus à propos que d'employer, au lieu d'éléments irrespirables, de l'air pur qui, après avoir servi à transmettre la force, ventilerait la galerie.

"L'honneur de cette importante application de l'air comprimé appartient aux trois ingénieurs Sommeiller, Grandis, et Grattoni, qui conçurent ensemble cette idée pendant une mission dont ils furent chargés en Belgique et en Angleterre."[2]

Nothing, as my readers know, is more difficult than to proportion exactly the respective merit due to inventors who have had the same object in view, and have adopted somewhat similar means for effecting the same.

I am particular in mentioning this invention of Mr. Bartlett's, because Mr. Brassey, with his usual generosity to those he employed, gave Mr. Bartlett £5,000 in aid of the expenses connected with the construction and trial of this machine.

While the Victor Emmanuel Railway was being constructed, Mr. Brassey and Mr. Henfrey contracted for making an extension of the Novara Railway from that city to the west bank of the Ticino at Buffalora, and also for the construction of the Chivasso and Ivrea Railway.

"It will be seen," as Mr. Henfrey says, "that the railways completed by Mr. Brassey and his partners formed a continuous line from the then French frontier at Culoz, on the Rhone, to the old Austrian frontier at Buffalora on the

[1] Paris, 1871.
[2] For a full and lucid discussion of the relative merits of Bartlett's and Sommeiller's Boring Machines, see the report of the debate at the Institution of Civil Engineers, Feb. 16, 1864.

Ticino, with the exception only of the pass over the Mont Cenis; and the years during which these contracts were executed comprised that bright period in the history of Italy, during which the kingdom of Sardinia, emerging from comparative obscurity, took its place by the side of the great Powers of Europe.

"Looking at the chain of events, we may reasonably speculate as to whether the facility for the movement of troops and supplies afforded by the railway communication for the whole distance from Paris to the Austrian frontier, excepting only the pass of the Mont Cenis, was the weakest argument or inducement brought forward by Cavour in soliciting Napoleon's aid for his country.

"We may speculate also as to whether these railways would have been so opportunely completed without the aid of British enterprise, at a time when it was most required; and perhaps be justified in concluding, that our countrymen may thus have borne a humble part in bringing about the greatest result of modern civilization—the unification of Italy."

CHAPTER XIV.

THE GRAND TRUNK RAILWAY OF CANADA.

(A.D. 1852-1859.)

Messrs. Peto, Brassey, and Betts.—Division of the work.—Mr. James Hodges.—Railways in America.—Bogie engines.—Mr. Rowan.—The steam excavator.—Wages to Canadians.

THE Grand Trunk Railway of Canada was one of the most important undertakings in which Mr. Brassey was ever concerned. This railway supplies a means of intercommunication through the valley of the St. Lawrence during the whole of the year, an advantage, which, owing to the river being frozen over for at least six months annually, had previously been enjoyed only during the summer. Even during the season, when the navigation is open, the means of transport, by water, are imperfect. Sea-going vessels, of 700 to 800 tons burden, could proceed safely as far as Lake Ontario; but the limited dimensions of the Welland Canal made it necessary that the produce from Lakes Erie, Huron, Michigan, and Superior should be conveyed to Lake Ontario in smaller vessels, not exceeding 300 tons burden. The Grand Trunk Railway was intended to obviate the necessity for this transshipment of cargo.

The first conception of this vast undertaking is due to the Honourable Francis Hincks, and the Honourable John Ross, who was for some time the Speaker of the Canadian House of Assembly.

During the summer of 1852, at the request of the Provincial Government of Canada, Messrs. Peto, Brassey, and Betts undertook an examination of that country, with a view to the development of a complete system of railways.

The execution of this task was entrusted to Sir William Jackson (who was afterwards associated with the contractors in their undertaking), and to Mr. Alexander Ross as Civil Engineer.

With the information thus obtained, a complete scheme for the Grand Trunk system of railways, including the Victoria Bridge, was prepared and introduced to the public under the auspices of Mr. Thomas Baring, and Mr. George Carr Glyn, the agents in England for Canada.

Mr. Robert Stephenson subsequently acted as Consulting Engineer to the Company, Mr. Alexander Ross being the Company's Engineer for the whole undertaking. Mr. Ross designed all the important "works of art;" the rest of the engineering being done by the contractors under him: the agents carrying out their work without any superior control. The railway was divided into four districts, the agents in command of these districts being on an equal footing with one another. They were in the habit of having consultations; but were not placed under the authority of any engineer-in-chief. Mr. Rowan, professionally educated as a civil engineer, Mr. Hodges, Mr. Reikie, and Mr. Tait, were the respective agents.

Hereafter, some account will be given of the extent and nature of some of the works on this remarkable line, which is one that forms a most important link in the system of American railways, opening up large districts of valuable land, and connecting the Erie and Great Western of Canada Railroads, and other lines of lesser importance. But, previously, it will be desirable to give a description of the nature and qualifications of the workmen in the great American continent, and their implements, as contrasted with the British workmen and their appliances.

This account will be chiefly taken from evidence furnished by the principal agents employed by Messrs. Peto, Brassey, and Betts in the construction of the Grand Trunk Railway.

The person upon whom I rely mostly for information in the present instance, is Mr. James Hodges. He began life as a carpenter, and was apprenticed to a carpenter and builder in Brompton. In 1853 he went to Canada to aid in carrying out the works of the Grand Trunk Railway,

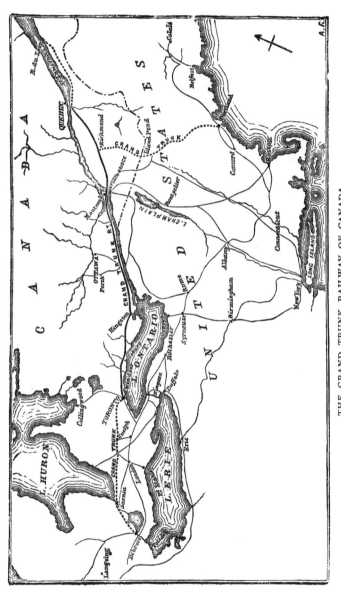

THE GRAND TRUNK RAILWAY OF CANADA.
—— Part made by English contractors ; .... part leased or made by Canadian contractors.

and remained there until the Prince of Wales had laid the last stone and put in the last rivet. Mr. Betts had the entire supervision of the local management of the line; but Mr. Brassey went over to Canada to see the progress, as was his usual custom in all works in which he was concerned, and also took part in the financial operations.

Mr. Brassey landed at New York, and Mr. Hodges went to meet him. The Americans showed the greatest attention to Mr. Brassey. Special cars were attached at the end of the trains for him, in order that he might have the best opportunities of seeing the country through which he passed; and the managers of the various lines always went with him.

As might be expected, he was greatly struck, and much delighted, with the new things which he saw in America. He was especially pleased with the granaries, and with the processes for cooling the grain adopted there, which are thus described:—

> By means of a thermometer, which is thrust into the middle of all large masses of grain, the Americans have the means of ascertaining its temperature. Be it ever so large a mass they find out directly it is heating. They then let it run down a long shoot, perhaps for 100 feet. It is then elevated again; and as it passes down rapidly through the air, it cools; and this operation also helps to brighten the surface of the grain. It is then in a cool state stored again, being returned back by means of little tin buckets which are driven continuously by steam engines. This system is now frequently used.

What, however, doubtless had most interest for Mr. Brassey must have been the works in America similar to his own in England; and herein he must have been struck by the contrast between the American Railway system and the English. Mr. Hodges justly says:—

"In America, a railway is like a river, and is regarded as the natural channel of civilization. Extended into a thinly populated district, it is the pioneer of civilization; it precedes population; and is laid down, even before common roads are thought of. As the expectation of traffic is, in many instances, but small, the cost of construction must be kept down as much as possible. With this object in view, timber is universally substituted for the more costly materials made use of in this country. Tressel bridges take the place of stone viaducts, and, in places in which in this country you would see a solid embankment, in America a light structure is often substituted."

In order to facilitate the construction of railways, the American Government has sometimes reserved a belt of territory, perhaps a mile in width, half of which is granted to the promoters of the railway. In South America the same principle has been adopted, and even carried further by the Argentine Republic. That Government gives the whole of this belt to the railway company, with the exception of certain portions of land near existing towns, or in places where it anticipates that great towns will hereafter be built.

Of course it would be impossible that such a principle should be adopted in a country like Great Britain, which is so thickly populated, and where land is so valuable.

Mr. Hodges dwells much upon the very ingenious and successful modes of facilitating labour by machinery, in use among the Americans, and adopted by them on account of the scarcity and consequent dearness of labour. He also notices the exceptional skill which the ironmasters have attained in that country and in Canada, being stimulated thereto by the difficulties arising from the distances which separate the iron mines from the beds of coal.

Many articles (he says) are made of cast iron in America, which in this country can only be produced in wrought iron. For example, cast iron wheels are made in America of a very superior quality to any which can be procured in this country. Here, cast iron wheels cannot be made to stand the same wear and tear as those made by hand: whereas in America, cast iron wheels are made to endure a very considerable amount of wear and tear, and the leading wheels of the locomotives, as well as all the wheels of the railway carriages, are constructed of cast iron; but the railway authorities in this country would not sanction its use for these purposes. In America, cast iron wheels are made of chilled iron, and they are found to answer their purpose admirably.

The marvellous ingenuity displayed by the Americans in the construction of all light machinery strikes the intelligent observer very much, and was, no doubt, thoroughly appreciated by Mr. Brassey. The greater part of the rolling stock for the Grand Trunk Railway of Canada had to be constructed at Birkenhead. For this purpose workshops were established; but previously to commencing any of the work, the contractors of the railway sent two clever mechanics throughout the United States, to examine the principal establishments in which similar railway stock was constructed. It must have struck everybody who has had any intimate acquaintance with America, how thoroughly fearless individual Americans often are of competition; notwithstanding that as a nation, they do not exhibit a similar fearlessness. These two mechanics were welcomed wherever they went; were supplied with drawings of all the best machines, and with every information requisite for their instruction.

It is to be recollected that this was in the year 1853: and at that time there were no such things made in England as morticing or planing machines, both of which are now so common. These machines were supplied by American manufacturers to the workshops of the "Canada Works" for the Grand Trunk Railway. It is some comfort to our national vanity to find that Mr. Hodges, after giving due praise to the American manufacturers for the lighter kind of machinery, adds:—

"I am bound to state, on the other hand, that in the construction of machinery for undertaking heavy work, England has carried the palm against every other nation.

"For this description of work Mr. Whitworth has given tools to the world which no other maker could have produced."

On the Grand Trunk Railway the contractors were obliged to adopt the American system of locomotives: these American engines were all constructed with "bogies." It is well known that "the bogie carriage yields to every irregularity in the railroad, whether it be horizontal or lateral, whereas, in an ordinary English locomotive, the leading wheels would soon be worn out from the violent oscillation and vibration arising from the traffic on the rough railroads which exist in America."

It may be noted that Mr. Hodges justly claims the "bogie" as an English invention; and he adds, "in real truth most of their (American) 'inventions' are English, which they have adopted."

His remarks with respect to the relative merits of English, American, and German labourers, are very valuable as coming from one who has had such large experience in the employment of labour. He speaks of the notable fact that, many of the most ingenious English mechanics were addicted to drinking; and he comments upon the injurious effects of Trades Unions. And here I must again give his own words:—

In England, the Trades Union dominates everywhere, and it soon ruins a man. He can only do just what its rules prescribe, and what he is allowed to do, and it is only one particular sort of work that he does. But in America, on the contrary, he is obliged to do all sorts of things. I am ready to admit that if a man only makes the head of a pin, no man can do that so well as the man who does that alone; but his intellect becomes cramped by that fact, and he soon loses all grasp of mind: but when he gets to the States he has perhaps to chop down a tree, or extemporise a pail, and that makes a man altogether different:—his intellect becomes clearer:—although he is not so good a machine, he rapidly rises in the scale of intellect. In England he is a machine, but as soon as he gets out to the United States he becomes an intellectual being. I do not think that a German is a better man than an Englishman; but I draw this distinction between them, that when the German leaves school he begins to educate himself, but the Englishman does not, for, as soon as he casts off the shackles of school, he learns nothing more, unless he is forced to, and if he is forced to do it, he will then beat the German. An Englishman acts well when he is put under compulsion by circumstances.

In executing the works that had to be effected on the Grand Trunk Railway, it was to be expected that the engineers, the contractors, and the agents should be led to consider the various contrasts to be found between American and Canadian work, and British work. It was the first time in the history of railways that those who had to conduct British enterprise abroad, found themselves in contact with men of their own race. Mr. Rowan's reflections upon this matter are very interesting and valuable. In giving his evidence upon the Grand Trunk Railway, he was questioned minutely upon various points connected with the relation of British workmanship to foreign workmanship generally, and especially of that part of workmanship which relates to invention. It was put to him that as labour is much more costly in America than in England, a stimulus was given to American manufacturers to produce articles which involved the least employment of labour. To this he entirely assented. It was then suggested to him that technical instruction is more developed abroad than in England, and his reply was as follows:—" My conviction is this, that we always in England excel the continental producers in the manufacture of any material that has once been established; but I think that improvements are most likely to originate on the continent from their greater theoretical knowledge. I believe that they possess much higher theoretical knowledge than we do, but there they stop. When a new invention or improvement has been established, and comes out of the dominion of rigid theory into that of practice, then I am of opinion that the Englishman always beats the foreigner."

As an instance of an invention which has been largely used in America, on account of the scarcity of labour there, Mr. Rowan mentions the Steam Excavator, and he says that:

Towards the last, in consequence of the extreme cost of labour, we employed steam excavators, not because they were cheaper than men, but because they supplied the want of labour, and enabled us to get on faster. A steam excavator is found to be profitable only in very hard material, such as hard pan, in which a very large force is required to excavate. In lighter materials, such as sand or gravel, it is more expensive to use than men at five or six shillings a day. We used them notwithstanding, even in filling ballast, and I undertook a large quantity of ballasting myself in that way.

This scarcity of labour gave rise to great difficulty in the execution of the railway works on the Grand Trunk Line. Wages were very high. A man who received five shillings in England per diem, would receive seven shillings and sixpence in Canada. This difference in the rate of wages was caused not only by the scarcity of labour, but by the circumstance that, in Canada, out-of-door work is impossible for four months in the year.

When Mr. Brassey went over to Canada to inspect the works, he suggested that they should endeavour to bring up a large body of French Canadians from Lower Canada. This suggestion was carried into effect by Mr. Rowan. A large number of Lower Canadians were brought up in organized gangs, each having an Englishman or an American as their leader. These gangers received a guinea a week for each man they brought. The French Canadians, however, except for very light work, were almost useless. They had not physical strength for anything like heavy work.

They could ballast, but they could not excavate. They could not even ballast as the English navvy does, continuously working at "filling" for the whole day. The only way in which they could be worked was by allowing them to fill the wagons, and then ride out with the ballast train to the place where the ballast was tipped, giving them an opportunity of resting. Then the empty wagons went back again to be filled; and so, alternately resting during the work, in that way, they did very much more. They could work fast for ten minutes and they were "done." This was not through idleness, but physical weakness. They are small men, and they are a class who are not well fed. They live entirely on vegetable food, and they scarcely ever taste meat.

These men, however, though their powers of work were but feeble, proved to be of great use, inasmuch as their coming prevented the stalwart men from leaving.

Mr. Brassey's main object in going to Canada was a financial one. The Canadian Government had lent three millions of money to the Grand Trunk Company; and these three millions had a priority of interest over all other claims upon the shares. Mr. Brassey succeeded in persuading the Canadian Government to remit the priority of their claims, which proved a great assistance to the Company.

In considering the difficulties which attend any railway enterprise, the first thing to be noticed is the nature of the ground through which the railway has to pass. Another difficulty, however, may be occasioned by the nature of the adjacent country. In the construction of the Grand Trunk Railway of Canada, this second difficulty must have been very great, when it is remembered that the conditions of the country through which the railway had to pass, were such, that a third of it passed through cultivated ground, and the other two thirds through forests. It may easily be imagined what difficulties this created in the way of housing men, procuring provisions, and bringing these provisions within reach for daily consumption.

CHAPTER XV.

GRAND TRUNK RAILWAY (CONTINUED).

(A.D. 1854-1860.)

The Victoria Bridge.—The ice bridge.—A council of Indian chiefs.—Floating dams.—Mr. Chaffey's steam "traveller."—Description and opening of the bridge.

THE following account of the difficulties which had to be encountered in constructing the Victoria Bridge, of how these difficulties were surmounted, and of the successful issue of the undertaking as a great work of construction, has, with very little alteration and abridgment on my part, been taken from a paper written by Mr. Hodges. It will be duly valued by the reader, as a succinct and complete narrative of the construction of one of the most admirable works which have ever been accomplished by British skill and enterprise in our Colonies.

The difficulties connected with the construction of the Victoria Bridge, and the doubts entertained as to its practicability, even after the inauguration of the Company, induced the Board of Directors to ask Mr. Robert Stephenson to examine and report upon the design, which he did in the summer of 1853, visiting Canada for that purpose. The structure, as it at present exists, was carried out from the designs of Mr. Stephenson and Mr. Ross, and under their joint responsibility.

"The site of the bridge is at the lower end of a small lake, called La Prairie Basin, which is situated about one mile above the entrance to the canal, at the west end of Montreal Harbour. At this point the Saint Lawrence is 8,660 feet from shore to shore, or nearly a mile and three quarters wide. The most serious difficulty in the construction of the Victoria Bridge, arose from the accumulation of the ice in the winter months. Ice begins to form in the Saint Lawrence in December. Thin ice first appears in quiet places, where the current is least felt. As winter advances, 'anchor' or ground ice comes down the stream in vast quantities. This anchor ice appears in rapid currents, and attaches itself to the rocks in the bed of the river, in the form of a spongy substance. Immense quantities accumulate in an inconceivably short time, increasing until the mass is several feet thick. A very slight thaw, even that produced by a bright sunshine at noon, disengages this mass, when, rising to the surface, it passes down the river with the current. This species of ice appears to grow only in the vicinity of rapids, or where the water has become aërated by the rapidity of the current. Anchor ice sometimes accumulates at the foot of the rapids in such quantities as to form a bar across the river, some miles in extent, keeping the water several feet above the ordinary level.

"The accumulation of ice continues for several weeks, until the river is quite full. This causes a general rising of the water, until large masses float, and moving farther down the river, unite with accumulations previously grounded, and thus form another barrier; "packing" in places to a height of twenty or thirty feet.

"As the winter advances, the lakes become frozen over. The ice then ceases to come down, and the water, in the river, gradually subsides, till it finds its ordinary winter level, which is some twelve feet above its height in summer. The 'ice bridge,' or solid field of ice across the river, becomes formed for the winter early in January. By the middle of March the sun becomes very powerful at midday, and the warm heavy rains rot the ice. The ice, when it becomes thus weakened, is easily broken up by the winds, particularly at those parts of the lakes where, from the great depth of water, they are not completely frozen over. This ice, coming down over the rapids, chokes up the channels again, and causes a rise of the river, as in early winter.

"In order to avoid the dangers consequent on these operations of nature, the stone piers of the Victoria Bridge were placed at wide intervals apart; each pier being of the most substantial character, and having a large wedge-shaped cut-water of stone work, slanting towards the current and presenting an angle to the advancing ice sufficient to separate and fracture it, as it rises against the piers. The piers of the bridge were in fact designed to answer the double purpose of carrying the tubes, and of resisting the pressure of the ice. In each of these respects they have fully answered the important objects sought to be attained.

"It was the duty of Mr. Hodges, as the agent of Messrs. Peto, Brassey, and Betts, to find suitable stone for the Victoria Bridge. The best stone was found on lands in possession of the Indians, and it became necessary to enter into negotiations with the chiefs of the tribe to whom the land belonged.

"A conference was arranged to take place on a Sunday, after church, that being the only time when a number of them could be brought together, sufficient for the transaction of so important a business. The assembled chiefs, thirteen in number, were not arrayed in paint and feathers, after the manner of Cooper's heroes, but were miserable, dirty-looking old men, with long hair, and they all smoked short clay pipes. At first they were disinclined to treat with Mr. Hodges, on the ground of his *extreme youth*. But, upon being assured that he was not less than forty, their objections were overcome, and no further difficulties were experienced in the conduct of the negotiations.

"The stone quarries thus obtained were situated at Point Saint Claire, sixteen miles west of Montreal, and about half a mile from the track of the Grand Trunk Railway. The stone is a very hard limestone, and, when exposed to the atmosphere, becomes of a light grey colour.

"The Saint Lawrence, where it is crossed by the Victoria Bridge, varies from five to fifteen feet in depth during the summer, and its bed is of limestone rock, with large boulders on the surface. This led to the contrivance of floating dams, which were warped into position, and scuttled immediately upon the opening of the navigation, and which were pumped out and taken back to a place of safety before the ice came down. These "caissons" were

188 feet in length, and 90 feet in width. Their bows were wedge-shaped, to stem the current, and the stern was made so that it could be removed, when the masonry was complete, thus enabling the floating dam to be shifted to various positions.

"The first caisson was towed to its position on May 24, 1854.

"The first working season at the Victoria Bridge was a period of difficulty, trouble and disaster. The agents of the contractors had no experience of the climate. There were numerous strikes among the workmen. The cholera committed dreadful ravages in the neighbourhood. In one case, out of a gang of two hundred men, sixty were sick at one time, many of whom ultimately died. After the harvest, towards September, the cholera at length disappeared; labour became more plentiful; and the work in consequence proceeded more satisfactorily.

"In the year 1855 great difficulties were experienced from the financial state of the Company, and the rise in the value of money, caused by the Russian War; but the works were nevertheless prosecuted with much spirit; and the abutment on the south side of the river was commenced.

"Before leaving England for Canada, Mr. Hodges made a sketch and description of a steam-traveller. One of the most eminent firms in England was consulted and employed to accomplish what was required, and, after two years of experiments and an expenditure of some thousands of pounds, a machine was sent out, which could never be made to do very much more than move itself about; and which, after various fruitless attempts to make it available, was thrown aside, and never used afterwards. In the meantime the same descriptions and drawings were shown to Mr. Chaffey, who was one of the sub-contractors, and had been in Canada a sufficient length of time to free his genius from the cramped ideas of early life; and during the winter of 1854 and 1855, the rough, ugly, but invaluable machine was constructed, and in the subsequent spring was in full work.

"This celebrated steam-traveller, as constructed by Mr. Chaffey, was erected at the junction of the Champlain

Railway with the temporary track for the bridge works. By it the whole of the stone for the works on the south side of the river was shifted. This traveller had a span of sixty feet, and ran along rails supported on gawntrees,[1] 1,300 feet in length and 50 feet in height, between which the stone was sorted and stacked ready for work. This machine unloaded the wagons, and stacked with the greatest ease the largest blocks of stone, some of which weighed ten tons. Over 70,000 tons of stone were twice moved by this machine. Only one man was required upon the traveller, while one other could stack the stone.

"And here it may not be out of place to observe again concerning the emigrant mechanic, who in his new home so often has to perform work without either proper materials or appliances, and is so often driven to contrive simple labour-saving machinery, how superior he becomes to the man he was when he first left home. It is curious to remark how a plodding man of this description, shut out, as he is usually considered to be, from all means of gaining information or knowledge, will become, in a short space of time, self-reliant, competent, and able. He may have scarcely any means or appliances at his disposal to accomplish that which, a few years before when at home, with everything at hand, he would after repeated attempts have abandoned as impracticable.

"This is only one illustration amongst many which could be adduced showing the development of the talents and skill of the members of Mr. Brassey's staff; men who, when they left home, gave little evidence of being above the ordinary mark, but who, in a foreign land, in difficulties proved themselves full of enterprise and resource.

"The shortness of the working season in Canada involved much loss of time. It was seldom that the setting of the masonry was fairly commenced before the middle of August, and it was quite certain that all work must cease at the end of November.

"Notwithstanding these disadvantages, the amount of material used in 1856, was, Masonry, 829,120 cubic feet; puddle clay, 13,223 cubic yards; timber, 344,450 cubic feet.

[1] A kind of permanent timber scaffolding.

"In June, 1858, it was decided that a great effort should be made to complete the bridge by the end of 1859. Every individual concerned put forth his utmost exertions, and on August 12, 1859, the foundation-stone of the last pier of the Victoria Bridge was laid, in the presence of some few spectators; and it is interesting to note that in just six weeks and two days from the date of commencing that pier, 108,000 cubic feet of masonry were laid, the whole being completed by September 26, 1859.

"It is impossible to describe all the difficulties experienced in the prosecution of the works: such as the 'shoving' of the ice at the commencement, and at the breaking up of the frosts; from the collision between floating rafts of vast dimensions, in some cases 250 feet in length, and the temporary staging, erected for the purpose of putting together the tubes; and from innumerable other causes. The whole of the iron work for the tubes was prepared at the Canada Works, Birkenhead—an establishment erected by Messrs. Peto, Brassey, and Betts, expressly for the manufacture of the bridge work and rolling stock for their Canadian contracts. At these works every plate was finished ready to be fitted into its proper place; and I must draw attention to the extraordinary perfection obliged to be attained in the preparation of this ironwork. I am informed that in the centre tube, consisting of no less than 10,309 pieces, in which nearly half a million of holes were punched, not one plate required alteration, neither was there a hole punched wrong! The importance of this accuracy may be estimated, on considering that, had any portion been carelessly prepared or even wrongly marked, a failure might have been the result, involving the delay of a year in opening the bridge, and consequently, a loss of many thousands of pounds. Great credit is therefore due to Mr. George Harrison, the manager of the Birkenhead Works, and to his able assistants, Mr. Alexander and Mr. Heap, for the successful completion of their share of the task.

"The following details and quantities of the materials used at the Victoria Bridge, together with the numbers of the men and horses employed, may be found interesting:—

Total length of the tubes, 6,512 feet.
Weight of iron in the tubes, 9,044 tons.
Number of rivets in the tubes, 1,540,000.
Number of spans, 25; viz. 24 from 242 to 247 feet each, one 330 feet.
Quantity of masonry in piers and abutments, 2,713,095 cubic feet.
Quantity of timber in temporary works, 2,280,000 cubic feet.
The force employed in construction included 6 steam boats, and 75 barges, representing together 12,000 tons, and 450 horse power.
3,040 men.
144 horses.
4 locomotive engines.

"At length, all difficulties having been successfully overcome, the bridge, of which the first stone was laid on July 20, 1854, was completed, as had been intended, in 1859.

"The bridge was first opened for the passage of trains on December 19, 1859, and the formal inauguration by His Royal Highness the Prince of Wales, took place in the following year.

"The devotion and energy of the large numbers of workmen employed can hardly be praised too highly. Once brought into proper discipline, they worked as we alone can work against difficulties. They have left behind them in Canada an imperishable monument of British skill, pluck, science, and perseverance in this bridge, which they not only designed, but constructed."

After warmly commending the staff of superintendents by whom he was assisted in constructing the works, and the workmen who laboured under their directions, Mr. Hodges observes that the successful accomplishment of his great task was primarily due to the spirit and inspiration he derived from the confidence reposed in him by his employers. "Looking back," he says, "at all the various difficulties, practical and financial, by which the work was from time to time embarrassed, it scarcely admits of a doubt that, in the hands of other and less energetic and persevering contractors than Messrs. Peto, Brassey, and Betts, it would not have reached the successful issue to which it has been brought. Amid every discouragement they stood stoutly to their task; and when the hearts of all around them seemed to fail, their encouragement, enterprise, and assuring confidence, kept everything going." As Mr.

Robert Stephenson enunciated, in the course of an address on the subject of the bridge at a dinner given to him in 1853 by the engineering profession of Canada, at Toronto, "the contractors left even the engineers themselves little more than the poetry of engineering."

CHAPTER XVI.

THE CRIMEAN RAILWAY, ETC.

(A.D. 1852-65.)

Sir Morton Peto's account.—Railway stores.—Mr. Beattie.—The Victoria Docks, London.—Thames Graving Docks.—Mid-Level Sewer.—Steam cranes.—East London Railway.—Danish Railways.—English and Danish education.

SOME of the minor works which occupied Mr. Brassey's attention during that very busy period of his life, from 1852 to 1865, deserve especial notice. When I say "minor works," I only mean that they were not works of such length and magnitude as the Great Northern line, or the Grand Trunk Railway of Canada. Almost every one of these minor works, however, possesses a peculiar interest of its own, and is therefore worth recording.

The first of them, and not the least remarkable, is the Crimean Railway. It is not only noticeable as being one of the greatest feats in railway making that has ever been known, but also as showing what is likely to be done in times of great danger for this country by those who are called the "captains of industry." The action of Government may be found on sudden emergencies to be rather stiff and constrained. That unwise habit of distrust, which seems to be creeping into the management of Government departments, may exercise large sway. The individual action of those departments will thereby have been crippled, and the fear of incurring responsibility increased. The result of neglecting the advantage of personal selection in the choice of agents, may, at a time of crisis, be found to be as injurious as the opponents of the present system imagine it will be. An economy, intense in small matters, and slack in great concerns, may have been adopted by the nation, and by that time have done its worst. Still, however, as long as we have men in the position of Mr. Brassey, and partaking of his nature, we shall have some persons who may, on a sudden emergency, come to the aid of Government, and play for a time its part, until it recovers itself, and is refitted for vigorous action.

I prefer to give in Sir Morton Peto's own words, addressed to Mr. Thomas Brassey, the account of what was done by himself, Mr. Brassey, and Mr. Betts, in this remarkable undertaking. Sir Morton says, "When I undertook this work in 1854, at the request of the late Duke of Newcastle, on the part of the Government, before doing so I took your father's advice, and he strongly urged it on me as a public duty, and promised me his aid in every possible way, and the whole was carried out at cost price, without any profit being charged.

"The organization of the detail and transport—a most laborious duty—was undertaken by my late lamented brother-in-law and partner, Mr. Betts, the general direction and administration resting with myself.

"We saw your father daily at this time. He advised on all the points, and helped by every means in his power, and I should not do his and Mr. Betts' memory justice, if I did not state that to them fully as much as, if not more than to myself, is the credit of the execution of the work really due.

"Our exertions were seconded by every railway company —the directors opening their stores for our free use at cost price.

"We succeeded in sending out twenty-three large steamers with men, horses, railway engines, commissariat and other stores, in a very short space of time, and within the first twelve days of the arrival of the first convoy we laid seven miles of line; and the soldiers handing shot and shell to each other were superseded to that extent in that time; and before the completion of the siege thirty-nine and a quarter miles of line were laid to every part of the front, and seventeen locomotives engaged in the conveyance of stores, &c. &c. I received a letter from Field-Marshal Burgoyne, then General Burgoyne, on his return from the command of the Engineering Staff, stating it was impossible to overrate the services rendered by the railway, or its effect in shortening the time of the siege and alleviating the fatigues and sufferings of the troops.

"I shall not do right if I do not refer to the admirable services of Mr. Beattie, our chief agent, who for three weeks after the commencement of the works never retired one night to rest, and died in four weeks after his return to England, from the effects of his devotion to duty in his engagements in the Crimea."

The next work I shall mention is the Victoria Docks, London, for the making of which Messrs. Brassey, Peto, and Betts contracted. "These were all carried out," Sir M. Peto says, "by Mr. W. Hartland for the firm, under the direction of Mr. G. P. Bidder, and were opened to the public in 1857. They are entered from the Thames, immediately below Blackwall, by a lock, having a depth of water on the cill of twenty-six feet at Trinity high water. They have a water area of over a hundred acres, divided by eighty-feet gates into a tidal basin of about twenty acres, and a wet dock of about eighty acres. They have vaults for wines, and warehouses for general merchandise, to the extent of about twenty-five acres of floor.

"The City warehouses for wines and general goods are in Fenchurch Street, and were constructed with a floor area of about five acres. They are in direct railway communication with the Docks. All the warehouses, quays, dock gates, &c., are supplied with, and worked by Sir William Armstrong's hydraulic machinery, and are connected with all the principal railways of the kingdom.

"The same firm constructed the Thames Graving Docks (Edwin Clark's patent), with an entrance from the wet dock of the Victoria Dock, having a water area of about fifteen acres, and hydraulic machinery and lifts for docking and under-docking vessels of all capacities."

Then there was the Northern Mid-Level Sewer, which Mr. Brassey contracted with the Metropolitan Board of Works to make. This, though only twelve miles in length, was a work of great magnitude and difficulty, and occupied nearly three years in construction. The line of sewer runs from Kensal Green, passing under the Bayswater Road, Oxford Street, and Clerkenwell, to the River Lea. Mr. Brassey took the greatest interest in this undertaking, which is, by some persons, considered one of the most difficult works that have ever been done in this country. It was necessary to tunnel under houses and streets, and also to cross the Metropolitan Railway with a very large tube. This work is noticeable because, for the first time, I believe, cranes worked by steam were generally employed in sewers for hoisting the earth from excavations direct into the carts in which it was to be removed. The method hitherto adopted had been that of lifting the earth from stage to stage by manual labour. The adoption of the new system enabled the contractors to dispense with a great deal of this manual labour, which is always found to be more costly in London than elsewhere, and it also accelerated the execution of the works.

Another work of small extent, but of great labour and difficulty, was the East London Railway, running from New Cross, through the Thames Tunnel, to Wapping. The difficulty in this construction was caused by the depth which had to be attained in order to arrive at the Tunnel. A tunneled approach was rendered unavoidable by the nature of the ground. Until the main drainage was carried into effect, the water used to rise up to the surface in great quantities. This railway was also a work in which Mr. Brassey took a very great interest. Indeed, wherever there was a difficulty in any of his works, there he was sure to be found. His brother-in-law, Mr. Henry Harrison, who was engaged in the construction of this railway, speaks of Mr. Brassey's fearlessness when examining the works. "He would walk without the slightest misgiving along a plank only twelve inches wide, over a chasm fifty feet deep."

Among the minor enterprises, though it was an enterprise of no light kind, the construction of railways in Denmark must be mentioned. Five hundred miles of railway were made by Messrs. Peto, Brassey, and Betts in Denmark, extending from Rendsburg and Tönning to

Aalborg in Jutland, with branches across the island of Funen, and across different parts of Schleswig. This system of railways also crosses Jutland to the town of Holstebro.

These Danish railways were constructed slowly, for the contractors were only allowed by the Government to do a certain amount of work annually—averaging in value generally from £200,000 to £300,000. It consequently took ten years to complete this system of railways. The work was mainly executed by Danes, for though English sub-contractors were at first taken out to Denmark, they were soon got rid of, as it was found that the cheapness of rye whisky, namely, about one shilling a gallon, was too great a temptation for them, and destroyed their powers of working.

The account that Mr. Rowan (the agent of Messrs. Peto, Brassey, and Betts) gives of the powers and habits of the Danes of all classes who were employed in these railways, is very interesting, and such, in fact, as rendered it desirable not to omit the mention of these works in Denmark, which otherwise were not of high interest.

When the English sub-contractors were got rid of, Danish sub-contractors were taken on. The Danes were found to be a very steady, and altogether very superior class of men. Mr. Rowan gives similar praise to the Danish common labourers. It must be observed, however, that they take their time to do their work in, beginning in summer at 4 o'clock in the morning, and not leaving off until 8 o'clock in the evening. How different is this mode of working from that of the English navvy, who, as we have seen, will sometimes get through an immense amount of labour early in the afternoon, lifting during the day nearly twenty tons weight of earth on a shovel over his head into a wagon. The Danish labourers have five intervals of rest in the day, these intervals lasting each for half-an-hour.

In Denmark rent is cheap; and the food of the common labourer, chiefly consisting of black bread, is also cheap; but fuel and clothing are dear. The labouring classes, however, in Denmark, dress much better than the same classes of workmen in this country.

Mr. Rowan notices that the Swedish workmen drank more than the Danes. They were energetic and polite, but troublesome; in short, they were not as steady as the Danes. It is gratifying to find that, after all, the British navvy is the king of labourers. Mr. Rowan evidently was greatly pleased with his Danish labourers; but when pressed with the question whether a Danish workman surpasses a British navvy, he replied in these words:—"No man is equal to a British navvy; but the Dane, from his steady, constant labour, is a good workman; and a first-class one will do nearly as much work in a day as an Englishman."

In the mere construction of the Danish railways there was not much which need be noticed. Occasionally there were very heavy and troublesome earthworks, so troublesome, indeed, that Mr. Brassey was sent for, to be consulted as to how the work should be dealt with; but there was not anything of that very special character that demands to be recorded in the history of railway enterprise. The character of the people, and the effect that their education has upon their character and conduct, are the main points worthy of notice in the story of these Danish works. Mr. Rowan had great opportunities of observation in reference to these particulars. The superintendence of the railways was entirely in the hands of military engineers, who were all trained in the Polytechnic school of Copenhagen. They had the very highest theoretical education; but, as Mr. Rowan says, they possessed no practical knowledge whatever on leaving these schools. The same statement may surely be made about the students in all schools, whether Danish or British. This remark, however, does not furnish a complete reply to Mr. Rowan's observations; and the further statements which he makes give a good illustration of one of the most important points to be attended to in education. "The great fault of Danish technical education is the overdoing of it. The young men are kept in school till they are twenty-five. They come out highly educated; utterly ignorant of the world, but educated to a tremendous height." The main point in which Mr. Rowan found that these highly educated persons were deficient, was decisiveness. "They have been in the habit of applying to one of their masters for everything, finding out nothing for themselves; and the consequence is, that they are children, and they cannot form a judgment. It is the same in the North of Germany; the great difficulty is, that you cannot get them to come to a decision. They want always to enquire and to investigate, and they never come to a result."

The foregoing is a very important statement. There is great reason for thinking that of all the qualities which are needful for the wise conduct of human life, decisiveness is the one which can least afford to lie dormant. It soon dies away by inanition, if not exercised. Moreover, it is very questionable whether it can be revived. Experience seems to show that if young people are not trained to decide, or at least not encouraged and allowed to exercise decisiveness, they will never be able to evoke this quality in after life when it is wanted. If this be so, it is a matter which requires the most serious consideration at the present time, when the British nation, or rather its Government, is introducing a system tending to promote the laborious acquisition of knowledge at an early period of life; which system, however, as the opponents to it contend, may produce a well-instructed and docile, but, at the same time, an unthoughtful, unoriginal, and indecisive race of men. I make no excuse for this short digression, any question relating to Education being of so much significance at the present moment. I may add, that I believe that our success hitherto in colonization, which has far exceeded that of any other people in the world, has greatly arisen from the fact of our possessing more decisiveness than those other people, and from our education having been less stimulated by material rewards, so that our youth have been accustomed to think a little for themselves, without being induced to turn their thinking at once into profitable courses.

Mr. Rowan gives an anecdote which serves well to illustrate his preceding statements:—"I was some time ago speaking to a man of business in Denmark, who is an exception to Danes generally, and extremely energetic and a man of great powers, besides being altogether a man who would make a first-rate man of business in England. I said, " My friend, will you tell me why it is you are so different from all your countrymen?" "Yes," he said, " because I learned my business in Liverpool." I said, " Will you go further, and tell me where is the difference between Liverpool and Copenhagen?" He said, " I will tell you in one word. If I had been learning my business here, and I came to a difficulty, I should straight go to one of my superiors in the office, and he would take a great deal of trouble to tell me how to get over the difficulty, and show me how it should be done. But when I was in an office in Liverpool, and I came to a difficulty, and went to my superior there, and asked him to explain it to me, he said, " Do not bother me about it; find it out!" " And," he added, "that is the secret."

The constructors of railways in Denmark had, as might be expected, an evil time of it during the war. The Danish Government seized the rolling stock of the company. The sub-contractors were obliged to assist in making military earthworks. The railway banks were formed into regular fortifications, and had to suffer bombardment. Indeed, as Mr. Rowan observes, " the combatants on either side had no compunction in seizing our materials, and in making our people work for them."

The only further evidence given by Mr. Rowan which need be alluded to on this subject, is evidence of a similar character to that so often given before as regards the confidence placed in their agents by these great contractors. The estimates for the Danish work were agreed upon by Mr. Rowan and the Government agents in detail; and the figures were never objected to when they were sent home to Messrs. Peto, Brassey, and Betts. " I do not know," Mr. Rowan says, " of a single instance, in which one of their agents has failed in that respect (in respect of misusing the confidence reposed in them): they prove themselves worthy of the trust, and that shows what there is to be gained by placing confidence in others."

CHAPTER XVII.

WORKS IN AUSTRALIA.

(A.D. 1859-1863.)

Mr. Wilcox.—Australian prices.—Emigration.

THERE are some men, the interest in whose lives, when these come to be recorded, lies wholly in the results of their daily work. There are other men in the record of whose lives the daily work they did is of no account with posterity, and may be summed up by the biographer in a few careless sentences. It is in vain that Charles Lamb, apostrophising the India House, exclaims: "Thou dreary pile, fit mansion for a Gresham or a Whittington of old. Stately house of Merchants, with thy labyrinthine passages and light-excluding pent-up offices, where candles for one half the year supplied the place of the sun's light; unhealthy contributor to my weal, stern fosterer of my living, farewell! In thee remain, and not in the obscure collection of some wandering bookseller, 'my works!' There let them rest, as I do from my labours, piled on thy massy shelves, more MSS. in folio than ever Aquinas left, and full as useful! My mantle I bequeath among ye."

It is in vain, I say, that this great humourist endeavours to persuade us that these were his works, which we know were not his works, at any rate for us; and we for ever refer to the obscure collection of some wandering bookseller for our knowledge of that much-suffering, most gentle, and most loving soul.

On the other hand, there are some men in whose career thought and action are so happily blended, at any rate happily for the biographer, that their daily labours form an admirable thread to the main narrative of their biography.

These are great warriors, statesmen, conquerors, and discoverers of new lands.

Again, there are lives in which the interest centres in great works done, not exactly of a continuous character, not exactly affording a good thread for biographical narrative, but of which it may be said, in the forcible word, that adorns and illustrates the so-called "monument" of Sir Christopher Wren; *Circumspice*.

This was eminently the case with Mr. Brassey's life. The hundreds and thousands of persons who are daily passing over railways, constructed by his energy, ability, and perseverance, might well, when looking at many a remarkable and difficult construction on these lines, and speaking of his merits, exclaim, *Circumspice!*

I am now going to treat of Mr. Brassey's work in Australia. There is one point of interest which it lacks—namely, that it was not subjected to his personal inspection or supervision.

But, on the other hand, there is a very important reason why the work in Australia should be brought before the reader; and that is, because it bears closely upon the great subject of Emigration—a subject which must have the deepest interest for all thoughtful men in this overpopulated country, which will yet have to consider the whole question of Emigration, with far more care than it has hitherto bestowed. Before entering into the details which have been furnished to me of the construction of these Australian railways, I would remark, that the evidence which we get about Emigration in this indirect manner from persons who have gone to Australia, with a purpose entirely foreign to the general subject of Emigration, is likely to be most valuable evidence. It will be full of knowledge, and yet it is not probable that it will be based upon any pre-conceived opinions regarding Emigration, or that it will have any personal bias regarding the interests of emigrants.

I feel that I am not deviating improperly from the main subject of this work in following out these indirect consequences of the late Mr. Brassey's labours. Nothing, I am confident, would have more delighted that good man, than to find that his work, in distant countries, had directly promoted the welfare of the native people amongst whom that work was accomplished, and that it had developed special information indirectly bearing upon the future welfare of his countrymen.

I now proceed to give the result of the evidence of one of the principal persons employed by Mr. Brassey in the construction of Australian railways. This gentleman's name is Mr. Samuel Wilcox. Before going to Australia, as one of Mr. Brassey's agents, he had been employed under Mr. Ballard on the Great Northern line, and also in Holland, on the line between Utrecht and Rotterdam. He had also been employed in the construction of the Paris and Caen Railway. During his employment on these lines he had enjoyed ample experience of the way in which Mr. Brassey dealt with his agents, and with all the persons, from the highest to the lowest, acting under him. In March, 1859, he went to Australia in company with Mr. Rhodes, another esteemed agent of Mr. Brassey's. The lines they had to construct were in New South Wales, and were called the Great Southern, the Great Northern, and the Great Western Railways. There is nothing in the construction of these lines for which it is needful to claim the attention of the reader. I mean that there were no remarkable engineering difficulties which had not often been surmounted in Mr. Brassey's previous undertakings. But the cost of the labour deserves to be carefully noted. Taking, for instance, any twenty miles of the Southern line as an illustration, Mr. Wilcox states, that if he had to lay down a similar length of line in England, upon English terms, the difference of expense in favour of England would be £3,000 or £4,000 a mile; in both cases exclusive of the rolling stock.

The system of organization was the same in Australia as it had been in England, namely, that of sub-contracts for all parts of the work. The iron work, the rolling stock, and plant of all kinds, came from England, but not, of course, the timber. It will now be interesting to see the rate of wages. Labourers earned from 7s. to 8s. per day, and at piece work would make 9s. A mason averaged 12s., a bricklayer received from 11s. to 12s., and a carpenter would earn from 10s. to 12s.

Now comes the question of food and its expense. Here I prefer to give the witness's own words:—

Q. Have you considered the cost of living?

A. Yes; a man would live uncommonly well there for about 8*s.* or 9*s.* a week individually.

Q. Suppose he had a wife and a family of four children; what would it cost them?

A. I can hardly tell you that.

Q. Take a man spending 10*s.* a week there; if he had been living in England would it have cost him 8*s.*?

A. He would get as much bread and meat there as he could eat, but here he could hardly look at it. As long as a man with a family is kept from drink there, he can, in a very short time, get sufficient money to start and buy a piece of land, and become "settled."

Q. May it not be said that a good stout labourer in England could not live as a navvy for less than 8*s.* a week?

A. Not living as a navvy does. I do not think that he could live on 8*s.* a week; living generously as a navvy has to live. Out there he could live very much more amply supplied at 10*s.*, and really on less than 10*s.* In the case of some of the men I have known camping out together, the rations did not come to more than 8*s.* 6*d.* per week.

Q. Did you find that a working man, placed as he appears to be in Australia in exceptionally advantageous positions with regards to means, drinks more?

A. Yes; he does.

Q. In short, there is a great deal of drunkenness there?

A. Yes; and the drink is more expensive; they charge you more there; they charge you 6*d.* for a glass of beer, and they charge for a bottle of beer 2*s.* 6*d.*, which you get for 1*s.* in England.

This is what may be expected; but it is very vexatious to find that the great advantage which the English labourer gains in Australia, from the increased rate of wages and from the comparative cheapness of living, is counteracted by his disposition to spend more money in drink; and that the result shown by Mr. Wilcox's evidence is that a working man in Australia, having greater means at his command, does drink more than a labouring man in England.

Mr. Brassey's agents found that it was desirable to get labourers from Great Britain; and their efforts are thus described:—

"Mr. Brassey wrote to Mr. Milroy, and got him to select a lot of men in Scotland. Mr. Harrison selected some, and Mr. Ballard took a great interest in the matter, and also picked out a lot: altogether from England and Scotland we got 2,000 men. We had to provide them with an outfit, in accordance with the regulations of the Government. The cost to us, for selection and outfit, averaged about £5, and the cost to the Government, for the passage, about £12 in addition: the cost of each man, therefore, was £17, or £34,000 altogether."

It is worthy of notice that Mr. Brassey's agents did not attempt to get back from the men the amount of money that had been paid for passage-money. "We sacrificed that," Mr. Wilcox says, "to get the men there. Having men in the country, we knew that they must work for somebody; and we also knew that we were in a position to pay them as much as, or more than, any one else. They were at liberty, on landing, to go where they liked; and some few, not a great number, but some few, never came to the works at all; but we found that we got a great part of them, and more came out by other ships."

At the time this evidence was taken, it was put to Mr. Wilcox whether, as an emigration agent for the time, he was not in a better position, than the authorities of a parish in England, which might wish to promote emigration. He admitted that he was, and for two reasons:— First, because the parish has no employment to offer the emigrant when he gets out to Australia; and, secondly, because the parish wants to get rid of the worst men; whereas he, as Mr. Brassey's agent, wanted to get hold of the best men. His views, however, on this part of the subject, must be held to have reference rather to the benefit of the colony than of the home country.

The subject was then discussed in reference to the individual emigrant; and though the witness admitted that a man who did not succeed in England would not succeed in Australia, his conclusion, after all, came to this, "that the worse kind of man could not contrive to starve in the new country," and that "there is nothing like pauperism in Australia."

He was then pressed with a question; whether, taking it for granted that the least successful working man would not become a pauper in Australia, it would be a safe venture to send out say 20,000 people a year to that colony. His answer was most distinctly in the affirmative. The labour market would absorb them all; and emigration might take place, on a very much larger scale than has hitherto been attempted, without overstocking the labour market; but he added this judicious proviso, "That the emigrants must *not* be sent out in too large batches, in order that they may be got away, up country, without being compelled to linger unemployed at the port of entry. If a vessel went to New South Wales and to Queensland every fortnight, with 400 or 500 emigrants, they could be taken away without difficulty."

This witness also confirmed the evidence we have had before from other persons, "that work at the 'diggings' did not, upon the average, yield the same rate of wages as other employment."

The foregoing evidence must be admitted to be very valuable as regards the general question of Emigration. The other point which I wish to insist upon is, that which I have before alluded to regarding the almost unlimited trust which Mr. Brassey placed in his agents. At that remote distance he could not well supervise or control the estimates, and certainly he did not attempt to do so. When asked, " Did Mr. Brassey look over your figures ? " the witness replied, " No, I had to take the work before I consulted him." Q. " He was absolutely in your hands ? " A. " He was indeed."

The final questions put by Mr. Brassey's son to this witness brought forth answers which may be instructive to all employers of labour :—

Q. Did your correspondence with my father produce upon your mind the impression that you were labouring with and for a watchful employer?

A. I do not think it was so much that, as the extreme confidence he always placed in his assistants. I think they could not help feeling that they had his confidence.

Q. You would say, I suppose, therefore, that the stimulus with you to exertion in Australia was rather the feeling that you enjoyed the unlimited confidence of Mr. Brassey, than that you were working under any kind of supervision?

A. Just so. I could not say too much in his favour.

CHAPTER XVIII.

THE ARGENTINE RAILWAY.

(A.D. 1864.)

The Rosario and Cordova Railway.—Mr. Woolcott's evidence.—Cordova.—Dr. Scrivener's account.

OF all Mr. Brassey's undertakings there is not one which has more attraction for me, and which, I think, will more interest my readers, than the Argentine Railway. I have for a long time thought that South America offers the finest opportunities for emigration and colonization. I have in another work,[1] given a description of the climate, the trees, the fruits and the cereals, which are to be found in some parts of that magnificent continent. I subjoin this extract:—

"It has lakes, rivers, and woods; and in the character of its scenery much resembles an English park. It is rich in trees of every description—cedars, palms, balsams, aloes, cocoa-trees, walnut-trees, spice-trees, almonds, the cotton plant, the quinaquina that produces the Jesuits' bark, and another tree of which the inner bark is so delicate and white that it can be used as writing-paper. There is also the ceyba-tree, which yields a soft woolly substance, of which the natives make their pillows.

"The fruits of this most fertile land are oranges, citrons, lemons, the American pear, apples, peaches, plums, figs, and olives. The bees find here their special home; and twelve different species of them are enumerated, some of which form their nests in the trees in the shape of a vase. The woods are not like the silent forests of North America, but swarm with all kinds of birds, having every

[1] "Spanish Conquest in America," vol. iv.

variety of note and feather, from the soft colours of the wild dove to the gay plumage of the parrot; from the plaintive note of the nightingale to the dignified noise of those birds which are said to imitate the trumpet and the organ."

I think that this Argentine enterprise of Mr. Brassey's will have more important results than any other of his undertakings. There can be no doubt that the railways which Mr. Brassey and his various partners constructed in Great Britain, France, Italy, and Austria, would have been constructed, though not perhaps so quickly or so well, by other persons: but the Argentine Railway is an enterprise of a most peculiar nature, fraught, as I trust it will be, with the greatest results. It is the first time in the history of railway constructions that railway promoters have been great colonizers. Hitherto colonization has mostly been preceded by conquest; and that fact alone has placed immense difficulties in the way of judicious colonization. Conquest, indeed, is too favourable a word to describe the buccaneering expeditions which, from the earliest periods of the world's history, have led to the discovery and occupation of new lands.

The principal point to be noticed in the formation of this Argentine Railway, is the contract entered into by the Argentine Government with the promoters. That Government, in their concession, engaged to put the company in possession of one league of land on each side of the railway throughout its entire extent, commencing at the distance of four leagues from the stations of Rosario and Cordova, and one league from each of the towns, San Geronimo and Villa Nueva, subject to the condition of such lands being peopled. The contractors for the railway, Messrs. Brassey, Wythes, and Wheelwright, accepted from the railway company one-half of the above-mentioned lands in part payment of their contract price. There were no special difficulties in the construction of the Argentine Railway. Indeed, so easy was the ground that for part of the way the rails had only to be laid on the sleepers over the bare earth, a simple trench at either side being dug to carry off any surface drainage.

I doubt whether, in the history of railway enterprise,

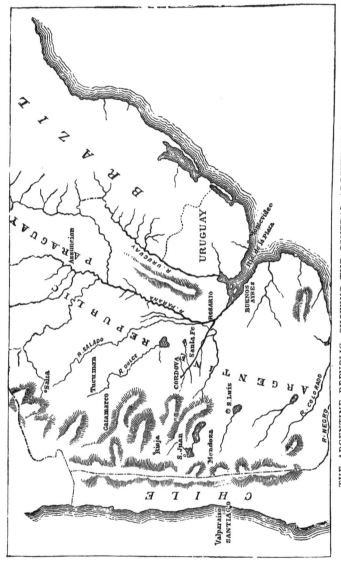

THE ARGENTINE REPUBLIC, SHOWING THE CENTRAL ARGENTINE RAILWAY.

there has been anything so largely beneficial to the country wherein a railway has been introduced, or anything which has afforded such favourable opportunities for emigration as this Argentine Railway—especially seeing that it is coupled with the possession by enterprising men, of land "marching" (to use a word well known in the north of England) with the whole length of the railway. It has been well observed that in most cases in which a new country is peopled, the colonization spreads out in somewhat of a circular or semicircular form. People, seeking new lands at cheaper prices, perpetually move onwards, throwing themselves and their fortunes into regions uncivilized, and unpeopled, or thinly occupied by hostile tribes. In going to the Argentine Republic, the colonizer may occupy land remote from cities, and therefore cheap, and yet find himself in immediate contact with one of the principal means and appliances of modern civilization. In ancient days, as we know, population and civilization followed the courses of rivers; but what are rivers as a means of transport, when compared with railways? Rivers are seldom navigable throughout their course, and are subject to great variations, affecting much their usefulness. He would have been thought a great magician in former ages, who could have promised, and have fulfilled his promise, to place for a great distance, through a most fertile but unpeopled country, something compared with which, as a means of transit, rivers, canals, and even Roman roads, sink into insignificance.

I think this matter so important and so intimately connected with my subject, as showing the benefits conferred by railway enterprise in general, and by this one of Mr. Brassey's enterprises in particular, that I shall subjoin some extracts from an account, with which I have been furnished, of the emigration into the Argentine Republic.

The Argentine Republic extends from 41° to 22° S.L., 1,320 miles in a straight line, The whole surface of the country is estimated at 726,000 square miles, being equal in extent to Russia. How stupidly, or at least how unfortunately, the world has hitherto been peopled! The population of Russia, that hard-featured country, is about 75,000,000: the population of the Argentine Republic, to which nature has been so bountiful, and in which she is so beautiful, is about 1,000,000.

"It is watered by the gigantic River Plate, the great rivers Parana and Uruguay, and a multitude of subordinate and navigable streams. Buenos Ayres, the commercial metropolis of the country, is a city of 200,000 inhabitants, of whom more than one-fourth are foreigners, and possesses an import and export trade amounting to nearly 700,000,000 dollars per annum.

"The tract of land conceded to the Central Argentine Railway Company lies on either side of their line, which connects the port of Rosario on the Parana with the city of Cordova, a distance of 246 miles, and comprises upwards of 1,200 square miles of some of the finest lands in the Republic. The tract lies between the parallels of latitude 31° and 33° south; the lands adjacent to Rosario being in 33°, and those nearest Cordova in 31°. They consequently lie within the temperate zone, with no great extremes of heat and cold. In summer the temperature in the open country is seldom above 80° in the shade, with cool nights; and in winter the average temperature is from 45° to 50° for two or three months, the mercury sometimes falling in the night to 32°, or freezing point. Snow never falls, but there are occasional hailstorms. There is no regular rainy season. Rain may be looked for at any time during the year, but it is generally more prevalent in the spring months of October and November, and the commencement of the winter season in May and June. Emigrants will observe that the seasons are reversed from those they witness in Europe. The winter season here is summer-time in the River Plate, and *vice versâ*. Fuel for the purposes of warmth is superfluous, but warm clothing and blankets are necessary in the winter months.

"Rosario, the starting-point of the Central Argentine Railway, is situated on the Parana River, one of the largest in the world, 180 miles above Buenos Ayres, and accessible at all seasons of the year for ships of 800 tons; with an excellent port, and facilities for loading and discharging very far superior to those possessed by the metropolis. A large number of ocean vessels, which are generally of 300, 400, and 500 tons register, resort to this port, taking cargoes of coal, lumber, railroad material, and merchandize of every description; and loading with wool, hides, flour, dried beef, tallow, hair, oil, bone-ash, copper, and other productions of the vast and rich territory that stretches for 1,000 and 1,200 miles to the west and north-west of Rosario. This town, which has only an existence of some fifteen years, and contains now a population of 25,000, is the natural outlet and seaport of the wealthy provinces of Cordova, Mendoza, San Juan, Tucuman, Salta, and others of less importance.

"Cordova, the central terminal point of the railroad, is situated in a deep valley on the banks of a river, amidst the most beautiful and varied scenery. Ascending from the city to the mountains, the traveller finds every variety of climate, with a difference of temperature at every additional ascent. It contains a population of about 35,000 inhabitants, and is the capital of a large and wealthy province, second only to that of Buenos Ayres.

"The province produces wheat, maize, and other cereals in abundance. Cordova wool obtains a higher price than any other in the market.

"Fruits of all kinds produced in temperate and semi-tropical climates are abundant—apples, pears, cherries, figs, grapes, pomegranates, oranges, lemons, &c. Strictly tropical fruit—such as pine-apples, bananas, and plantains —do not grow.

"The great range of mountains in the immediate vicinity of the city contains mines of copper and silver, and quarries of various kinds of marble.

"The best proofs of the adaptability of the Argentine Republic to agricultural pursuits are: the larger amount of wheat and corn already grown in Cordova and forwarded to Rosario for shipment; the increase of the wheat crop in the province of Santa Fé, within the last seven years, from 40,000 to 1,000,000 of bushels, with the corresponding increase in all other farming products; and the undeniable success and prosperity of the agricultural colonies in the country, although generally composed of second and third-rate immigrants.

"The only thing which, perhaps, may have a depressing effect on the spirits of a new comer on arriving to establish himself on the Argentine plains, is the absence of trees, and consequently of shade for himself and his cattle.[1] The fertility of the soil, its adaptability for immediate cultivation, the deliciousness of the climate, the grassy waving fields, are perhaps momentarily forgotten in the monotonous aspect of the pampa or prairie. Thus it is of importance that the colonist should surround himself, as soon as possible, with what Nature has forgotten to endow these regions—shade and fruit-trees.

" Dr. Scrivener, who has himself resided for many years in the country, says: ' The climate is fine and healthy; the lightness of the atmosphere produces an exhilarating effect, and an increase of energy and activity.' And, in alluding to the Andine and Cordova ranges, he remarks: ' I have traversed these mountains on many occasions, and am therefore enabled to form an opinion of the salubrity of the climate, as also of that on the route from the province of Cordova to the shores of the Pacific. All over this vast tract of land, that fatal enemy of man, the tubercular phthisis, so justly feared by the inhabitants of Lima and Buenos Ayres, is entirely unknown.'

" During the summer months field labourers can earn from 6s. 8d. to 8s. 3d. daily. Many of the wealthy and most prosperous men in the country have risen from the very lowest positions. 'Labouring men,' says Mr. W. Hadfield, ' have always done well in the country; and as they can save nearly all the wages they earn, a steady man can soon save up sufficient to purchase a share in a flock of sheep. The lowest wages for a country labouring man is about £2 a month; but as he becomes more accustomed to the labour and ways of the country, he gets to earn double and treble that sum in four or six months, being, of course, fed and lodged free of any expense to himself.

" ' The total imports in the five years, 1865-69, were of the average *annual* value of £6,540,000, while the exports

---

[1] This may appear inconsistent with what has been said at the beginning of the chapter, in reference to the beauty and variety of the trees. But the truth is, that the lands of the Company along the course of the railway are situated in the lowlands: the description of the trees applying to the higher regions, and has especial reference to those parts of the southern part of America which were occupied formerly by the Jesuits.

averaged £4,970,000. The railway lines open for traffic in September, 1870, were of a length of 458 English miles, while sixty miles more were under construction at the same date, 210 more miles were contracted for, and 400 miles were in course of being surveyed.'"

The foregoing extracts show the great capabilities for emigration opened up by the Argentine Railway. At present, the unsettled political state of many parts of South America must be admitted to be a drawback to emigration to that continent. This drawback, however, is one which will continually diminish by the gradual influx of emigrants from Europe. These will form a compact body, able to resist any incursions of the natives, and also able to control and master any difficulties that may arise from the troublous condition of certain of the neighbouring States. The want of good government has been the bane of many portions of a region of the world which is not only very fertile and very healthy, but which has also resources which make it likely to be one of the greatest centres of commercial enterprise.

CHAPTER XIX.

MOLDAVIAN RAILWAYS.

(A.D. 1858-64.)

Difficulties of negotiating.—Messrs. McClean and Stileman.—Prince Sapieha.—Marquis Salamanca.—Failure of the negotiation.

EVERY railway, or group of railways, the construction of which has been mentioned in this work, has been intended to illustrate some special circumstance of railway formation. The Moldavian Railways are now brought forward in order to show the difficulties of negotiation which often precede the construction of railways or any other public works.

In December, 1858, M. Adolphe de Herz, then of Frankfort, addressed Mr. Netlam Giles a letter, proposing the formation (through Mr. Brassey) of a company for constructing a railway from the Austrian Carl-Ludwig Railway, at Lemberg, in Galicia, to Czernowitz and the Bukowina frontier of Austria; and thence, through Moldavia, by Roman to Galatz on the Lower Danube, with branches to Jassy (the capital), and to the salt mines of Okna.

This railway was to be upwards of 500 miles long, and roughly estimated would cost about £6,500,000.

In the reply to M. Adolphe de Herz, delay was suggested on the ground that, in the face of the Emperor's speech to the Austrian Ambassador on the 1st instant, it was utterly impossible to hope that capitalists would at that moment entertain the question of constructing Austrian railways.

It was urged upon M. de Herz that he should not press his project *now:*—"wait a few weeks:" it was said,—"we shall either have war or peace:—nothing can be worse than the present uncertainty:—no one will listen to you now; not even if you offer diamonds for chalk stones."

If it were not a too self-evident proposition, one might dilate upon the injury to all good work effected by war, or by the fear of war. The Emperor of France, on his fête day, makes a remark, which is not supposed to be friendly, to the Austrian Ambassador, and immediately a good work, for distant Moldavia, is set aside.

The war between France and Piedmont and Austria having terminated, the railway negotiations were re-commenced; but, as the conditions for a concession of the Moldavian section of the project had been agreed upon in favour of M. Mavrojeny of Jassy, and as no concession had then been demanded from the Austrian Government of the section from Lemberg to Czernowitz and the Bukowina frontier, it was decided that efforts should, in the first instance, be directed towards the construction of the Moldavian section (about 300 miles). Mr. Giles thereupon introduced the project to Messrs. McClean and Stileman, and to Mr. Brassey. Messrs. McClean and Stileman undertook the engineering, and as no surveys existed, Mr. McClean, in that enterprising and liberal spirit which all who knew him must recognise, as he, too, has been one of the foremost leaders of labour in our time, offered to report upon the line. Accordingly, in September, 1861, the whole line from Lemberg to Galatz was examined by Mr. McClean; and on November 25, Messrs. McClean and Stileman made a report, recommending the contract for the Moldavian section to be given to Mr. Brassey, Sir Morton Peto, and Mr. Betts at the sum of £2,880,000, or £9,600 per mile.

On April 25, 1862, a concession was granted by the United Principalities of Wallachia and Moldavia to M. Mavrojeny and the Prince Leo Sapieha (chairman of the Carl-Ludwig Company) of the Moldavian portion of the above railway (300 miles), with a guarantee of £6 per cent. on a capital fixed at £11,584 per mile.

There was, however, in this concession a condition which rendered it valueless. It was stipulated that the whole 300 miles should be completed in five years. The Government of the Principalities being only a year old, its credit in the markets of Europe was not such as to make it in the least degree probable that the requisite sum could be raised. Messrs. Glyn, Mr. Brassey and others proposed that the concession should be modified by dividing the line into sections, to be executed successively; and they offered to provide the funds, and to construct the first section from Galatz to Adjud (80 miles) upon the Government guarantee named above. But the Government refused to modify the concession. Accordingly, the project in its entirety was laid before the public at the end of June, 1862; but they were not attracted by the project, and did not subscribe the requisite capital.

Prince Sapieha having requested Mr. Brassey's opinion as to the best mode of proceeding, Mr. Brassey addressed to him the following letter:—

London: July 18, 1862.

Prince,—After full consideration of the Moldavian Railway project, it seems that we are both of opinion that there is a serious defect in it; namely, that it has no junction with your Carl-Ludwig Railway at Lemberg; and I fear you will have considerable difficulty in obtaining the support of the public to an isolated scheme for the Principality of Moldavia.

If a company could be formed for the entire line from Lemberg to Galatz, with the branches to Jassy and Okna, it would, I think, be favourably received; and I venture to suggest that your Highness endeavour to form a combination with Baron Anselm Rothschild and your friends at Vienna for carrying it out.

You will easily be able to form an approximate idea of the capital required; and should my co-operation as contractor be thought desirable, you may consider I will accept one-third of the contract price which may be agreed upon in shares of the company.

I shall be in Paris to-morrow night, and will make a point of conferring with Mr. Talabot on the subject.

I have the honour to be, &c.,
THOMAS BRASSEY.

Baron Anselm de Rothschild and M. Talabot declined to embark in the undertaking, and nothing was done in the matter until June, 1863. In that year Messrs. McClean and Stileman, again willing to facilitate the project, made definitive studies of part of the line at their own expense. After protracted negotiations at Bucharest, the promoters of the railway succeeded in making a preliminary arrangement with the Government for a new concession, with a guarantee of $7\frac{1}{4}$ per cent., on a capital fixed at £12,800 per mile, instead of 6 per cent. on £11,584 per mile as originally granted. The line to be constructed in independent sections, and the Principalities to subscribe one-fourth of the capital. These conditions were submitted by Mr. Brassey and the other promoters of the railway to the International Financial Society, in August, 1863, but they declined to co-operate in the undertaking.

Not daunted by these repeated failures or discouragements, Mr. Brassey and his friends in the winter of that same year, renewed the negotiations at Bucharest with the hope of obtaining a definitive concession in the terms agreed upon in the preceding June. Meanwhile, however, an adversary had entered the field, the well-known Spanish banker and capitalist, the Marquis Salamanca, who, with M. Gustave de la Hante, had offered to take the whole line, and relieve the Government from their subscription of one-fourth of the capital. The Government announced their preference of Señor Salamanca's offer to that of Mr. Brassey, and recommended the Chamber to accept it. Mr. Giles then said to the reigning Prince Couza, "Let Salamanca and De la Hante have the concession. I return to England, and wish your Highness good morning." The Prince, however, would not hear of this abrupt departure. What the Prince desired was a fusion between Salamanca's party and Mr. Brassey's. Eventually this was effected; and a concession granted for the whole line to Salamanca, De la Hante, Mavrojeny, Sapieha, Peto, Brassey, and Betts, upon terms which Mr. Brassey and his friends informed the Government at the time would prove unacceptable to the public.

Then there was a meeting between Mr. Brassey and Salamanca in London. The terms of the concession were, that the concessionaires might issue three-fourths of the capital (about £4,000,000) in bonds, Salamanca's view being that the line could be made entirely with the proceeds of the bonds, and the shares (whatever they were worth) would be the contractors' profit.

The Marquis proposed to issue the bonds at once; and Mr. Brassey said, "Mr. Salamanca, before we can issue bonds, the shares must be paid up: and I am not prepared to say that we can get these shares placed."

Several schemes were suggested and discussed for getting over this difficulty, none of which however were satisfactory to Mr. Brassey, who, with his characteristic scrupulousness, declined to assent to any course, except that of a *bonâ-fide* sale of the shares, or an advance upon them to the extent of the value they represented.

He therefore said to the Marquis, "Look here, Mr. Salamanca, if you and your friends will put £500,000 down on the table any day you like to name, I and my friends will do so too: then the shares will be paid up," and there can be no possible objection to the bonds being issued.

The view taken by the British contractor on this occasion is one which will certainly recommend itself to the public; and it affords a striking instance of the extreme sensitiveness of Mr. Brassey in those dealings in which the public were concerned. There were further negotiations between Mr. Brassey and Salamanca, but they came to nothing. Ultimately, the Marquis obtained another concession on his own account from the Government of the Principalities; but he was again unable to carry it out.

Mr. de Herz, the original proposer of the railway, again appears upon the scene. He had settled at Bucharest as the manager of the Roumanian Bank, and had discussed the question of the railway with Prince Couza, who was greatly discouraged on account of the Marquis's efforts having failed. The Prince asked if Mr. Brassey would recommence negotiations—a question which was conveyed to the English promoters of the railway.

Meanwhile, the line from Lemberg to Czernowitz had been constructed, and Mr. Brassey was upon the point of completing the line from Czernowitz to the Moldavian frontier at Suczawa, on behalf of the Lemberg and Czernowitz Company. It was thought desirable that the concession should be demanded by this Company. The Chevalier d'Ofenheim therefore proceeded to Bucharest, and on June 7, 1868, a concession was granted to Mr. Brassey and others, nominees of the company, for that part of the original project which extended from the Austrian frontier to Roman, with branches to Jassy and Botoschani, under the terms of $7\frac{1}{2}$ per cent. upon £14,000 per mile, with a subvention (as *fonds perdus*) of £2,500 per mile in addition, terms nearly twice as onerous to the Government as those asked on the previous occasion.

Mr. Brassey was employed as contractor for the Moldavian lines comprised in the concession to the Lemberg and Czernowitz Company; and a portion of these lines, namely, those from Roman and Jassy, were completed and opened in 1870, in Mr. Brassey's lifetime.

Thus Mr. Brassey, after negotiations extending over ten years, completed only 360 out of the 500 miles of which the original project, mooted in 1858, consisted. The remaining 140 miles between Roman and Galatz were conceded to Dr. Strousberg in 1868, no part of which has yet been opened.

The story of this contract affords a notable instance of the quantity of work, in the way of negotiation, that Mr. Brassey and other great contractors have had to undertake before the commencement of their labours of construction.

CHAPTER XX.

THE INDIAN RAILWAYS.

(A.D. 1858-1865.)

Mr. Henfrey's evidence.—Eastern Bengal Railway.—Delhi Railway.

I AM not able to give any of those minute details respecting the construction of railways in India which have been given in narrating the construction of railways by Mr. Brassey and his partners in other regions of the world.

It would, however, be a very wrongful omission if I were to omit giving some account of these railways.

The account that I am enabled to produce is entirely in the words of Mr. Charles Henfrey. It is so clear and succinct that I have not ventured to alter or to abridge it; and though it consists only of a lucid statement of the principal facts, it will doubtless be read with interest by those who concern themselves with the affairs of that great empire in which the British race has shown more skill in government, more aptitude for organization, and, I think I may say, more consideration for the native peoples, than has been witnessed in any other part of our vast dominions and extensive colonies. Not the least boon that we have been the means of giving to the people of India, is the railway system that has been established there. In this small island, now intersected in all directions by railways, the advantages, however great, of railway communication are small when compared with those derived from the introduction of railways into countries of great extent and magnitude, such as Canada and India.

There are, for instance, many consequences likely to ensue from railway construction in India, which may be adduced to support the rather strong statement I have made in the preceding sentence. I may mention some of them; such as a probable breaking down of caste; the opportunity for bringing students from great distances to the central establishments of instruction; and, that which must be considered the greatest, the means afforded by railway transit of lessening the calamitous effects of those periodical famines which have, at various times, rendered desolate extensive regions in India. Mr. Brassey and his coadjutors must ever be held to take a high place among those who have furthered colonization in Canada, and material welfare of all kinds in India.

Mr. Henfrey's statement is as follows:—

"In 1858 Mr. Brassey, in partnership with Mr. Wythes and Sir Joseph Paxton, undertook the construction of the Eastern Bengal Railway, a line 112 miles in length, commencing at Calcutta, and terminating at a village named Kooshtea on the river Ganges.

"Mr. Brunel was the consulting engineer in England, and Mr. W. Purdon the chief engineer residing in India. The firm had been in treaty with Messrs. Hunt and Elmsly, who had just completed a contract on the East Indian Railway, for the execution of the Indian work of this contract; but, the negotiation falling through, they arranged with Mr. Henfrey, who had just completed the Ivrea Railway in Italy, to join them as a partner, and to go out to India in charge of the contract.

"Mr. Henfrey arrived in Calcutta in March, 1859, but was unable to make a serious commencement of the works until the following November, owing to the Government not having placed the contractors in possession of the land.

"Mr. Brassey's usual good fortune did not attend him in this enterprise. The contract had been entered into on the basis of a schedule of prices, to be applied to work performed; and these prices had been arrived at from Indian experience before the mutiny, the last embers of which had been stamped out in 1858. There was no other guide to Indian prices; for but little work had been carried on during the mutiny, and that little under very exceptional circumstances.

"In 1859 the great political changes that had taken place, and the extinction of the mutiny, gave an extraordinary impetus to public works in India, and the prices of labour and building materials rose upwards of 30 per cent.

" In consequence of the numerous works being carried on simultaneously, such as the East Indian Railway, the South-Eastern Railway, the drainage works of Calcutta, extensive alterations in the Calcutta Circular Canal, &c., great difficulty was experienced in obtaining a sufficient amount of labour for this contract, even at the enhanced rates.

"In spite of all these drawbacks, and in the face of a heavy loss on the Indian portion of the contract, the works were perseveringly pushed forward, and the line was completed, and opened throughout for traffic at the end of the rainy season of 1862.

" The cost of this railway, including rolling stock, was about £14,000 per mile.

" Mr. Brassey's next Indian contract was for the construction of the Delhi Railway; in partnership with Mr. Wythes and Mr. Henfrey.

" Mr. G. P. Bidder was the consulting engineer to the company in England, and Mr. Joseph Harrison the chief engineer resident in India.

" This railway, commencing at Ghazeeabad, a short distance from Delhi, on the East Indian Railway, terminated at Umritsir, in the Punjaub, the length being 304 miles.

" The firm undertook the construction of this line, including all the works, and providing permanent way, station materials, and partially rolling stock, at the fixed rate of £14,630 per mile.

"The works included some very long viaducts over the rivers Jumna, Sutlej, and Beeas, besides many minor structures over rivers which would have been thought important on any other railway.

" All the ironwork and machinery were imported from England, and had to be carried upwards of 1,000 miles from the ports where they were landed. Including rolling stock, these materials weighed nearly 100,000 tons.

" Mr. Henfrey, with a large staff of assistants, arrived

in India at the beginning of 1865, and was enabled to make a general commencement of the works after the rainy season of that year.

"Much less difficulty was experienced in obtaining labour on this contract than on the Eastern Bengal Railway, mainly owing to the Firm having become known to the natives, and to their having established a reputation for fair dealing and punctual payments.

"Besides the local labour of the Punjaub, a great number of work-people from Bengal, Oudh, and the North-West Provinces, flocked to the line so soon as it was known that the works were fairly commenced: and throughout the execution of the contract a sufficient amount of labour was at all times obtainable.

"The stipulated date for the completion of the line was the 3rd of May, 1870; but sections of the line were opened for traffic in anticipation of this date, and so early as November, 1868, the eastern half of the line, from Umballa to Ghazeeabad, was opened with becoming ceremony by the Viceroy Sir John Lawrence. The whole railway was completed within the contract time, with the exception of 7 miles across the Sutlej Valley. It had been found necessary in 1869 to lengthen the Sutlej Viaduct from three-quarters of a mile to one mile and a quarter. This additional half-mile of Viaduct was built and was available for traffic within twenty months from the order being given for it in India, so that in October, 1870, the communication between Delhi and Umritsir was uninterrupted."

CHAPTER XXI.

RECOLLECTIONS OF HIS SON.

Love of engineering.—Mont Cenis Tunnel.—"Great Eastern" and Fell Railway.—Isthmus of Darien Canal.—Chicago.—Love of nature.—The Maison Carrée.—Taste for Art.—Love of Yachts.—Hospitality.—Politics.—Correspondence.—Advice to Parents.—Dislike of reproof.

I HERE insert a chapter composed entirely of extracts from the correspondence I have had with Mr. Thomas Brassey,[1] the present Member for Hastings, on the subject of his father's life.

I have said before that I have always been entirely against the writing of eminent men's lives by their sons; for sons are manifestly prevented, by due respect and filial love, from commenting upon the faults of their fathers. Now these faults, which sons are naturally averse from alluding to, are often only exaggerations of merits, only indicate that want of proportion in character which is to be seen even in the most remarkable men—and perhaps indeed, more in them than in others; notably in very successful men, for great success is often the result of great disproportion in character.

It will be seen here that I have succeeded in eliciting from Mr. Brassey's son what, in his opinion, were the faults of his father's character. And it will also be seen that these faults were, as I suspected they would prove to be, exaggerations of merits—virtues carried in this world to an extreme. It could hardly be expected that a man such as Mr. Brassey—whose kindness of heart, and gentleness of nature, were manifestly developed to a great extent—should not sometimes be led into error by a superabundance of these admirable qualities.

I will no longer detain the reader from commencing a

[1] Now Lord Brassey.

chapter in which I am sure he will feel the greatest interest.
Mr. Thomas Brassey says :—

"The picture of my father would be incomplete, which should represent him as contracted in tastes and sympathies, and caring only for his business pursuits. True it is, that he often used to say, in reply to suggestions that he should allow himself more rest and recreation, that he found his greatest pleasure in the administration of his large undertakings; but it is not the less true that he took a lively interest in a wide and varied range of subjects.

"The very nature of his business was such, that it necessarily expanded the mind and quickened the powers of observation.

"He delighted in all that appertains to the art of the engineer. He felt a keen interest in the designs which he was employed in executing, and often discussed and criticised their scientific merit. Whenever, in his business journeys, he found himself in the neighbourhood of a great engineering work, he made a point of examining it with care. At one time he was frequently in Savoy. The opportunities thus afforded of closely watching the progress of the tunnel under Mont Cenis, recently completed, were eagerly seized. He was in the habit of studying with attention every novel project in the engineering world. The 'Great Eastern' steam ship; ocean telegraphy; Mr. Hawkshaw's project for a tunnel under the Straits of Dover; the Fell Railway; the electric telegraph, when first brought out as an invention of practical utility; and many other engineering novelties, attracted his close attention; and, in many cases, their development was largely due to his liberal pecuniary aid.

"Dr. Cullen's project for a canal through the Isthmus of Darien was closely examined by my father and Sir Charles Fox. At their expense a survey across the isthmus was made; and the hope was entertained that material assistance might be obtained from the Governments of France and England. Our parliamentary system, however, is not favourable to such enterprises; and the late Emperor of the French, though deeply interested in the idea and the plans for its development, did not think it prudent to take any active steps.

"When travelling abroad or at home my father was an attentive observer of the population, and of every indication of the commercial and agricultural resources of the country through which he was passing. His visit to Chicago—which city is perhaps the most conspicuous example of sudden prosperity, accompanied by a corresponding growth of the population, which the world has ever seen—excited his intense interest. On his return home, instead of dwelling upon the misfortunes which had been the occasion of his visit to Canada, and upon his disappointment at the result of his mission, he would dilate upon the innumerable contrivances of American ingenuity which he had admired so much at Chicago and elsewhere. He had an admirable faculty, perfected by constant exercise, for estimating the traffic which a railway in any given district would probably secure. A journey through Canada —a newly settled country, with an uncertain future— would naturally lead him to study all the elements of industrial prosperity which were presented to his observation. On his return to England the results of his observations were laid before those who were interested in the subject with admirable lucidity, and in the most compendious form.

"He felt a great delight in the beauties of nature. He was frequently induced to diverge for a day or two, when out on a business journey, to visit a beautiful country. Mountainous scenery afforded him immense pleasure. I can remember his invariable enjoyment when travelling in a hill country, whether in Wales or Scotland, or amid the yet grander scenery of Switzerland and the Tyrol.

"Whenever he travelled abroad, he was a busy sightseer. He used to visit the churches, the public buildings, the picture galleries, with the keenest interest. He would seldom leave a great city, though the primary object of the visit would probably have been some matter of business, without giving almost as much attention to its works of art and its architectural monuments as the ordinary traveller, whose only object is the love of art or change of scene.

"I remember, during my Rugby days, an agreeable

journey with him to the South of France: his object being to inspect the works on the Lyons and Avignon Railway, at that time under construction. After he had completed his examination of the line, he determined to devote a couple of days to an excursion from Avignon to Nismes. On our way from the station at Nismes to the hotel, we passed the Maison Carrée, so justly celebrated for the exquisite perfection of its architectural proportions. I do not think that he had heard much about this building, perhaps he might never have heard of it before; but he immediately appreciated its great beauty, and remained at least half an hour on the spot, in order that he might thoroughly examine that admirable monument of ancient art from every point of view. The excellent judgment in architectural art, and the sincere and unaffected enjoyment of the beautiful, which he displayed, in the instance to which I have referred, made a strong impression on my youthful mind.

"Since that time I have observed, on numberless occasions, the same judgment and the same love of fine architecture exemplified.

"He had a special appreciation of fine sculpture. I have often heard him say that he cared more for sculpture than painting. When he visited the Royal Academy Exhibitions, in the old quarters in Trafalgar Square, he never failed to dive into the dark and gloomy cave to which British sculpture was so unworthily consigned. Years ago he had been introduced to Gibson's meritorious pupil, Spence, the sculptor of that charming figure, the Highland Mary. Spence was a Liverpool man; and in the first instance, it may be, moved by old local sympathies, my father was a liberal patron of his art. Among the many works which Spence executed for him, the most important was a large group, the 'Parting of Hector and Andromache,' which occupied a place of honour in the International Exhibition of 1862. The sculpture galleries of the Louvre were a favourite resort; and the Venus of Milo was a joy to my father, as it must be to all who have a true perception of what is lovely in the sculptor's art.

"Though he may have had a stronger feeling for beauty of form than for the harmony of colour, the poetical composition, and the wider range of sentiment, which the painter is enabled to display, he had a great delight in fine paintings. Wherever he went, even in the busiest epoch of his career, he never neglected the opportunity of visiting important collections of pictures in the cities through which he passed. During the last two years of his life, when his broken health at times almost incapacitated him from devoting the entire day to hard work, he often found a grateful relaxation at Christie and Manson's, or in the tempting rooms of Messrs. Agnew.

"The same love of the beautiful, which made him delight so much in architecture, sculpture, and painting, attracted him to the less pretentious, but still very real merits, of well-designed furniture, tasteful dress, and handsomely decorated porcelain. I have seen him linger long, and with great interest, in the furniture courts of international exhibitions.

"For porcelain he had a taste of early development. Just in the same way as, when he travelled on the Continent, he would be sure to visit any important gallery of paintings, so he would, when travelling in his own country, visit its great factories and workshops, and all that it possessed of interest and importance. In my boyhood I was once his companion, on a visit of inspection to the North Staffordshire Railway. We spent a day at Stoke, and devoted almost the whole morning to Minton's pottery and porcelain works. My father followed attentively every stage of the manufacture; and specially admired the skill of the ill-paid artists, innocent perhaps of the value of their own talents, who, working in a room of small dimensions and oppressive atmosphere, are employed in painting upon the *porcelaine de luxe*, which adorns the tables of the wealthy, rich nosegays, luscious fruits, and sweet landscapes—things so well imagined, but so seldom seen, by men in their position. In later days he could seldom accomplish a visit to any of the great Exhibitions, in which British ceramic art has made so creditable a display, without indulging himself in some of the beautiful works which he had seen. Rarely were these purchases intended for himself. They were almost invariably bestowed upon his relatives and friends.

"The same universal appreciation of what is beautiful, in whatever form it may find expression, made him feel an interest even in yachts, which are my peculiar hobby. In 1851, long before I had ever dreamed of yachting, the celebrated 'America' came to England, and astonished our old-fashioned yacht-builders by her marvellous speed. My father happened to be at Portsmouth at the time, and felt so deeply interested in the 'America,' that he hired a boat —I remember well the evening on which it occurred—and made the boatman pull several times round the yacht, as she lay in Portsmouth harbour. Seen from ahead the sectional lines of the 'America' above water present a close resemblance to those of a duck. He instantly detected, and highly appreciated, this analogy to a form which an attentive study of Nature seemed to have suggested to the builder.

"Again, in 1858, when an immense fleet of yachts was assembled at Cherbourg, on the occasion of the opening of the new docks, and the visit of Her Majesty to the Emperor of the French, my father—who was present on the occasion officially, as contractor for the railway from Caen to Cherbourg, opened by the Emperor on his journey to Cherbourg—exhibited a lively interest even in the yachting features of the pageant. I was then, thanks to his generosity, the owner of a small yacht; and he spent hours in pulling to and fro in my gig, admiring the many beautiful vessels of the pleasure fleet of England anchored there. I remember that he noticed with especial commendation the fine appearance of the schooner 'Constance.' Most yachting men, who were at Cherbourg in 1858, will endorse his opinion.

"It may seem unimportant, and yet, as another illustration of the wide range of subjects in which a liberal mind may take delight, I think I ought to allude to the admiration which he was wont to express for the troopers of the Household Brigade, their stature, their horses, and the style in which they always turn out. I have seen him, after one of those handsome soldiers had passed by, turn round, and watch him striding along the street or the Park, with all the delight which the officer commanding one of Her Majesty's regiments might have felt himself.

"Though unable, from a variety of circumstances, to occupy a large house, or reside very regularly in London, he was always fond of gathering his friends together at the social board. On such occasions he was the kindest and most genial host. His buoyant spirits and warm welcome made every guest feel happy.

"Clever people, who, from want of a complete education, continued long enough to establish for the remainder of life the habit of study, are unable to gather instruction easily from reading, are generally able to apprehend the more readily whatever is imparted to them by oral communication. My father, who was no reader, was a fastidious and excellent critic of public speaking. To hear the addresses of our most gifted orators on great public occasions, was to him an exceeding delight.

"His enormous business demanded every moment of his time during the working day, and left him no leisure for literary culture or reading newspapers, still less for reading books; but by a kind of intuitive faculty, and by gleaning what could be gathered in conversation with those whom he saw from time to time during the day, he was always well posted up in the more important political events of the time.

"He very rarely read the reports of debates in Parliament; but I can remember, many years ago, the admiration which he always expressed for Mr. Disraeli's speeches, which amid the flood of oratory which constituted the debates in Parliament, he ever found time to read.

"He also greatly admired the speeches of the late Lord Derby—at that time Lord Stanley, and a member of the House of Commons.

"My father had the highest opinion of the value to the French people of the state of order, and the material prosperity secured to them by the Imperial régime.

"Without making any pretensions to being a politician, my father's proclivities were Conservative: nevertheless, thinking that he had no claim to advise others upon matters with which he did not conceive himself to be specially conversant, he never on any occasion offered me advice on political matters, or remonstrated with me on a single vote which I gave in the House of Commons. On the contrary, if he ever said anything on the subject, it was rather to suggest that it was my duty to attend regularly, and assist by my vote the Government which I had been elected to support.

"When I first entered upon the curriculum of a classical education, my father felt almost as unable to estimate the progress I was making, or to give me advice in the prosecution of my early studies, as the domestic hen, who views, with mingled anxiety and surprise, the brood of pheasants or ducklings, which she has tenderly reared, betake themselves to the wild life of the woodlands, or to their first navigation in the adjacent pond. But he was careful to commend to my attention whatever acquirements he himself could fully appreciate.

"Reading was an art which he specially advised me to study; and, on my return from school at the close of each half year, he never failed to test the progress which I had made, by asking me to read to him a chapter in the Bible.

"To hear a good reader was a real pleasure to him. I remember having been with him at an entertainment in Exeter Hall, when the celebrated Vandenhoff read some passages of Scripture, which were introduced between the recitatives and songs of an oratorio. The old man's style of reading was admirable, and my father long afterwards was accustomed to dilate on the excellence of his elocution.

"In the conduct of the immense and widely scattered affairs on which my father was engaged, he necessarily directed a large part of his operations with the pen. But his correspondence sometimes exceeded what was strictly necessary. Every man has a hobby. Correspondence was his hobby. Mainly from the pressure of business, partly also from his love of the work, every spare moment was given to his correspondence. Until his power of writing had been impaired by a stroke of paralysis, he never made use of a short-hand writer, and he wrote all his letters with his own hand.

"Years ago, when annually visiting Scotland for the purpose of shooting over the moors in Dumfries-shire, of which he had a lease jointly with Mr. Locke, it was amusing to see how a bag, containing writing materials and a budget of letters to be answered, always accompanied the luncheon-basket. After a short walk on the moor, he would screen himself behind a stone wall, or retire to some shepherd's hut, and there proceed to write his letters with the same method and diligence which he would have employed in an office in London.

"He never allowed a letter to remain unanswered; and, though his correspondents were frequently persons in the humblest walk of life, and a large number of the letters addressed to him consisted of solicitations for loans, or for the exercise of friendly influence, he always gave a kind reply to all those who addressed him.[1]

"My father, ever mindful of his own struggles and efforts in early life, evinced at all times the most anxious disposition to assist young men to enter upon a career in life. The small loans which he advanced for this purpose, and the innumerable letters which he wrote in the hope of obtaining for his young clients help or employment in other quarters, constitute a bright and most honourable feature in his life.

"His success in life induced many parents, who could not see how to start their sons in the world, to ask his advice; and, as usual, a disposition was shown to prefer a career which did not involve the apparent degradation of learning a trade practically, side by side, with operatives

[1] In regard to Mr. Brassey's letter-writing, Dr. Burnett remarks:—
"I recollect the first time we went to Switzerland, we had to wait a day at Lucerne for the arrival of Netlam Giles, with whom an appointment had been made to accompany him over the Grimsel, the question of a tunnel being at that time (A.D. 1851) mooted. We sailed up the lake to Floelen; and, having walked a good deal during the day, I retired to bed before nine o'clock, leaving Mr. Brassey in the coffee-room. The next morning, on coming down to breakfast, I found him with a packet of letters he had written the previous night, having sat up till past two. I counted thirty-one letters. I was curious to see to whom this amazing number were addressed, and I noticed most of them were to sub-contractors, and related to their work; several were in reply to applications for employment, and so on. I never remember his leaving a letter unanswered. He would carry his letters about with him in a sheet of paper, no portfolio, no memorandums; his memory was astonishing."

in a workshop. But my father, who had known by his wide experience the immense value of a technical knowledge of a trade or business, as compared with general educational advantages of the second order, and who knew how much more easy it is to earn a living as a skilful artisan than as a clerk possessing a mere general education, always urged those who sought his advice to begin by giving to their sons a practical knowledge of a trade.

" He was of a singularly patient disposition in dealing with all ordinary affairs of life. We know how, whenever a hitch or delay occurs in a railway journey, a great number of passengers become irritated almost to a kind of foolish frenzy. He always took these matters most patiently. He well knew that no persons are so anxious to avoid such detentions as the officials themselves, and never allowed himself to altercate with a helpless guard or distracted stationmaster.

" Without pretending to literary skill, he possessed a remarkable power of stating clearly the terms agreed upon after an elaborate or difficult negotiation, or of giving advice and instruction to his subordinates upon the technical details of his business. He would sometimes give directions upon the most minute arrangements, even as to the supply of a bellows for a blacksmith's forge. It would be an injustice were I not to add, that the same benignity and courtesy which marked his conduct in every relation of life, pervaded his whole correspondence.

" In the many volumes of his letters which are preserved, I venture, with confidence, to affirm that there is not the faintest indication of an ungenerous or unkindly sentiment; not a sentence which is not inspired by the spirit of equity and justice, and by universal charity to mankind.

" You have often asked me to point out faults in my father's character. If there be a blemish in one who was as free from fault as seems possible to frail human nature, I would say that hesitation to condemn openly those errors of others, of which he was perfectly sensible, and which inwardly he judged with the severity which they deserved, was one of the few defects of his character. An incapability of refusing a request, or rejecting a proposal, strongly urged by others, was a defect in his character, as a man of business, and the principal cause of the greatest disasters which he experienced. He seldom formed a wrong judgment upon the merits of any business proposed for his acceptance; but he was often induced by others to enter into engagements which he believed *ab initio* to involve excessive risk, or be fraught with disaster.

"Akin to this defect of character, there was another peculiarity. He would often approve and expressly commend that in others, which, if done by the members of his own family, he would disapprove and oppose. This remark would apply more especially to cases in which he feared that our actions or practices might be misinterpreted as an exhibition of pride. But this peculiarity should perhaps be regarded as a virtue rather than a fault. It originated in his anxious desire to avoid doing anything which could give offence to others. In every relation of life his conduct was marked by the utmost refinement of feeling, and by the true spirit of a gentleman. He never failed in consideration for the feelings and susceptibilities of others; showing deference, without being servile, to those, to whom deference was due; recognising the superior social position which they enjoy who possess the advantages of long lineage, and ancient and large landed estate; yet knowing that in his own busy and remarkable career there was something honourable to himself, and which it was a distinction to have achieved. He was graceful in every movement, always intelligent in observation, with an excellent command of language, and only here and there betrayed, by some slight provincialisms, in how small a degree he had, in early life, enjoyed the educational advantages of those with whom his high commercial position in later years placed him in constant communication; but these things are small in comparison with the greater points of character, by which he seemed to me to be distinguished. In all he said or did, he ever showed himself to be inspired by that chivalry of heart and mind, which most truly ennobles him who possesses it, and without which one cannot be a perfect gentleman."

CHAPTER XXII.

CLOSE OF MR. BRASSEY'S LIFE.

Fitness for his work.—Agents and their masters.—Mental refinement.—Appreciation of others.—Feminine nature.—Fell Railway.—Illness.—Venice.—Death.

THERE is a painful circumstance attendant on biography, which does not necessarily belong to any other form of narrative, not even to history. In history it often happens that the most important and interesting personages, though they quit the scene of history, still live on; perhaps in honoured retirement, at any rate in some retreat from their labours. And you have not to read of or to record their failing health, or their death.

But in biography you have almost always to narrate the gradual decay of power, and, ultimately, the death of one, to whom, even if you knew him not in life, you, the biographer, have become insensibly attached. For, I suppose, there is no instance of a biographer entirely resisting the natural impulse to become attached to the person, whatever may have been his faults, who is the subject of the biography.

How much more strongly must this feeling exist when the biographer, in a long course of investigation, with the secrets of a life laid open to him, finds no serious blot; and, as he proceeds in his work, becomes only more and more cognizant of the merits of his hero, and more and more conscious of his own inability to depict a beautiful character which has thus been intimately made known to him,—more intimately perhaps than the character of any of those with whom he lives in close and daily converse.

Such has been the effect upon me in writing this life of Mr. Brassey. All that I said at the commencement of the work in reference to Mr. Brassey's character has been amply confirmed by the evidence which has been submitted to me; and I have discovered new traits of character, all of which have been of the most pleasing kind.

I spoke, at the commencement, of Mr. Brassey's presence of mind, a quality which has been repeatedly manifested in the course of the narrative; and it is one which more almost than any other commands the confidence and the respect of agents and subordinates.

I mentioned his unwillingness to blame anyone, even when it was needful to do so. This also has been shown throughout the life.

The reader must have perceived how singularly well fitted Mr. Brassey was for the work which devolved upon him. He had almost every qualification which one's imagination conceives to be requisite for such work. Early in life he had mastered the details of nearly every kind of labour which it was necessary to understand for the accomplishment of great works of construction. But, as we have seen, he did not stop there. His self-education was of a higher kind than that which contents itself with the mastery of work in detail, however needful it may be to acquire such a mastery. He attained that most valuable art which belongs to the master rather than to the man—namely, of dealing with details in masses; of leaving minutiæ to those whose business it is to attend to such things; and of directing and supervising work, instead of doing it all himself. This great change of occupation is not often easily accomplished by men after their youth has passed, it being then a somewhat difficult matter to transform a subordinate into a principal.

He was eminently endowed with that rare gift, which all men have in them as a potentiality, and which all can recognize, but which so few can produce when wanted—namely, common sense. I have submitted to him matters not connected with his own department of affairs—putting before him what lawyers call "an A and B case"—and have found his judgment admirable in such matters.

The trustfulness which was mentioned in the first chapter as one of Mr. Brassey's merits, has been amply exhibited throughout this volume. Indeed, I have so often dwelt upon his trustfulness, that I think I ought to explain why, in my judgment, it is a characteristic of such importance.

I have often thought that the question of the relation of agents to principals is one of those which has not been adequately considered by mankind. No two men in the world have exactly the same way of doing anything. This is strikingly to be seen in compositions of all kinds. There are a number of facts to be set forth, and the conclusions therefrom to be stated. If two persons were appointed to do this work independently, and the result was that the facts and conclusions were fairly stated by both of them, it is still probable that there would not be a single sentence exactly similar, in the two written statements. Now this illustration is seldom to be met with in life, for the general rule is, that the subordinate drafts and the principal corrects. But it may be noticed, in the case of a wise principal, that he makes a just allowance for the inevitable difference between his way—his ideal way—of doing anything, and his subordinate's, and does not expect that the modes of expression of any two men will be exactly coincident, or even nearly similar.

Apply this to all kinds of work; and it will be found that the judicious master not only places the "wise confidence," I have spoken of above, in his agents; but is able to abstain from unwise interference and needless criticism, and to be content with allowing his work to be done by other people somewhat in their own way, so that it be well done.

Where most men fail in governing is, in not entrusting enough to those who have to act under them. Most human beings intend well, and try to do their best as agents and subordinates; and he is the great man, who succeeds, with the least possible change of agents and subordinates, in making the most of the ability which he has to direct and supervise. Besides, men must act according to their characters; and he who is prone to confide largely in others, will mostly gain an advantage in the general result of this confidence, which will far more than counteract any evil arising from that part of the confidence which is misplaced.

From all I have seen of Mr. Brassey's conduct, as a

principal, I am convinced that he was one of the most judicious masters as well as one of the kindest; and that, looking ever to results, he thoroughly understood the art of leaving his agents to do their work in their own way, when minute interference was needless—all interference, as he well knew, having a tendency to check an agent's energy and his power of reasonable assumption of responsible authority.

I have also been confirmed in the view which I originally took of what I conceived to be "the ruling passion" of his life. This was to execute great works which he believed to be of the highest utility to mankind; to become a celebrated man in so doing;—celebrated for faithfulness, punctuality, and completeness in the execution of his work; also, for this was a great point with him, to continue to give employment to all those persons who had early embarked with him in his great enterprises, not by any means forgetting the humbler class of labourers whom he had engaged in his service.

It was a remark made by her, who must have known Mr. Brassey best, that he was "a most unworldly man." All that I observed of him, and have heard of him, confirms this view of an important part of his character.

The foregoing remarks apply closely to what was said at the beginning of the work. The new traits which I have discovered may chiefly be enumerated under the heads of generosity and tenderness. I certainly did not suspect at the outset that Mr. Brassey was as generous a man as I have found him to be. A very busy man has not always time to practise generosity, and still less to evince tenderness. But throughout the investigations I have been obliged to make in the course of this work, I have discovered almost innumerable instances of his generosity. He was never wanting when there was an opportunity to be generous. The same delicacy which in his lifetime made him so thoroughly reserved as regards these matters, and prevented his ever alluding to any act of liberality on his part, also compels his son and myself to maintain a similar reticence. But the reader may be assured that where I have, here and there, given an instance of generosity, I might have adduced twenty similar instances.

With regard to his tenderness, this is also a difficult subject to dilate upon, but it furnished one of the deepest traits of his character, and, in fact, was one of the foundations of that character, if I may so express it. He was very tender-hearted in all the relations of life—to his family, his friends, his subordinates, to all who ever came in contact with him. When he committed what appeared to be an error in judgment, it was an error of heart, hardly ever of intellect. He could not bear to wound anybody's feelings. It pained him to refuse; and, altogether, his tender-heartedness must ever have been to him, as it always is for such men in this world, a source of much trouble, distress, and suffering.

Mr. Brassey was a man of much refinement of mind. This kind of refinement, strange to say, does not come by education, which can only produce a superficial and constrained semblance of it. Real refinement is innate; and, as I believe, for the most part, hereditary. Mr. Brassey possessed it in a high degree. No coarseness of thought, no coarseness of diction, however veiled, received anything but the most marked discouragement from him. My readers will appreciate the value of this quality in a man who had to be a ruler over numbers of his fellow-countrymen, and whose example in this respect, if a bad one, would have been largely imitated. This quality made itself felt throughout the various transactions in which he was engaged. It was not only that coarse and violent language was checked in his presence; but the pain he evinced at all unkind wrangling, and at the manifestation of petty jealousies, operated strongly in preventing their being displayed before him. As one, who was most intimate with him graphically observed, "his people seemed to enter into a higher atmosphere when they were in his presence," conscious no doubt of the intense dislike which he had of everything that was mean, petty, or contentious.

This kind of refinement is not exactly a quality which was imperatively needed in Mr. Brassey's vocation, or which we should have expected to find in a person so employed. It was, however, eminently serviceable to him; and I think it may be noticed that those persons succeed most thoroughly in any calling, who have some qualifications which do not appear needed for, or even remarkably suited to those callings. For instance, it may be observed, as I have ventured to remark before, that lawyers, clergymen, physicians, statesmen, diplomatists have often succeeded greatly in their respective vocations by reason of their possessing qualities which are somewhat adverse to those qualifications which are supposed to be, and which perhaps are, of the first necessity for success in a clerical, legal, medical, political, or diplomatic career. And I doubt not that the refinement, the tact and the courtesy which, from his earliest years, Mr. Brassey manifested, far from hindering, greatly assisted him in his important enterprises. Tact was a very noticeable quality in him. His secretary says, " he always seemed to say the right thing to people at the proper time."

Mr. Brassey was singularly free from vanity and boastfulness; and was never heard to refer to the difficulties he had overcome, the position he had attained, or the great works he had accomplished.

I would not have the reader think lightly of the high courtesy and exquisitely good manners of Mr. Brassey. Some say that the world is hardening and coarsening. I do not agree with the accusation contained in the former word; but am constrained to say there is some truth in the latter. Indeed, one might contend that the world is improved in almost everything but good manners. The polite races of mankind have met with a hard fate: some of them have almost vanished from the face of the earth, and others have either been subjugated, or have not been able to hold their own in the world. Moreover, speaking of individuals, politeness of late years has been little encouraged amongst them. Men have risen to high stations who are singularly destitute of the good manners which, in former ages, would have been considered necessary for the posts they occupy. Then, again, the hurry, the fussiness, and the want of time in our generation, have militated against the cultivation of good manners. Lastly, and this is not a slight point, a fancy has prevailed that courtesy and refined manners are often characteristics of a disingenuous mind. From all these causes, there has certainly been a decline of good manners in the present century. It is, therefore, singularly pleasing to find that a man who had a great deal of rough work to do in the world, and whom no one can accuse of want of good faith in anything he said or did, manifested, throughout his life, the highest courtesy to persons of all classes with whom he came in contact.

He was a brave man—brave in a business like way. He seemed to think that it was his duty, in supervising work, to do whatever a practised workman was accustomed to do. Even after he had had a stroke of paralysis, which affected one of his legs, causing him to drag it wearily after the other in walking, he would, to the dismay of those who accompanied him, walk across a narrow line of planks, over great heights, when a false step, or a momentary giddiness, would probably have been fatal to him.

There was one characteristic of Mr. Brassey, which, in its highest development, is very rare; and that is, that he was a most appreciative person of other people's merits and labours. This, too, not merely in endeavours that were cognate to his own; but he admired and loved good work done in any capacity, or directed to any worthy end. Good speaking, good acting, good reading, good painting, even good dressing and judicious entertainment, were all delights to him. If a thing was to be done at all, he wished it to be well done; and, when it was well done, he rejoiced in the well-doing, and admired the doers.

One of the principal charms in the character of women is that they are eminently appreciative. Let us hope that when they take a larger share—the share that justice and reason admit—in the active labours of the world, they will not lose any portion of this charm. The poet Coleridge maintained that the greatest men are those in whom something of the beauty of the feminine nature is conjoined with the sterner qualities of the masculine character. If this be the case, Mr. Brassey might certainly claim that element of greatness, for he had that tender and respectful regard for the distinguishing merits of his fellow men which especially belongs to the highest order of womankind.

In thus delineating Mr. Brassey, I have drawn a character of a somewhat perfect kind. I have been compelled to do so. I know well that more faith is given to the biographer when he is able to point out considerable errors and defects in the character of the person whose life he relates; and, therefore, as a biographer I might have been glad to have had some of these defects to set forth, as they would have insured a higher credence for the merits of the character. But, in truth, Mr. Brassey was in mind one of those happily constituted and well proportioned men, who show forth a certain completeness of nature. And really the only quality I have had to comment upon, which can detract from that completeness, was the amiable weakness which rendered it almost impossible for him to give a direct negative at once, or to say to anybody anything, however needful to be said, which should partake of the nature of blame. This delicacy of mind might have been ruinous to him, if, from other qualities, he had not attained to such eminence in his calling that he had, as it were, the command of the chief contract business that was going on in the world. Here and there misadventure might occur on account of his not being staunch enough in refusal, or too reluctant to condemn; but, upon the whole, these pardonable errors had very little effect upon the main result.

Mr. Brassey was very persuasive and conciliatory; and when these qualities are joined to a reasonable amount of pliability, they are very effective. The greatest difficulties in the conduct of business are not those which are inherent in the business itself, but those which arise from the prejudices, tempers, and vanities—especially the tempers—of the men who have to manage it. Complicated as human affairs are in the present state of civilization, the minds of men are more complicated still, and in their management lies the main element of successful administration.

Mr. Brassey was not only a very warm and affectionate friend, but he was extremely solicitous to have the approval of his intimate friends in all the works he undertook. This remark applies especially to Mr. Locke, the well-known engineer, who has been mentioned in these pages, the patron, fellow-labourer, and most intimate friend of Mr. Brassey's middle life, and to Mr. Wagstaff, his legal adviser and executor, and the congenial companion of his later years. To the latter gentleman he would write almost every day, and sometimes twice a day, telling him what he had done, seeking his approval and asking for his advice. This trait is very noticeable in a man who had so much clearness of resolve, and who might, perhaps, have been expected to be very indifferent to the judgment of other persons about his doings.

Of the service that Mr. Wagstaff was to Mr. Brassey, it is difficult to speak in sufficiently high terms. He was, if I may so express it, an *alter ego* to his friend; and those only whose lives are perpetually exercised in deciding day by day upon questions requiring great discretion, can fully estimate what it is to have a friend to whom daily reference may be made in all the difficulties that arise. It may also be observed, that those only gain such friends who deserve to have them—those who can, with supreme frankness, tell the daily story of their lives to another, and who are of that modest and confiding nature that they can receive and act upon advice which may even be most unwelcome to them, as in some manner condemnatory of their own views and prepossessions.

This biography is now drawing to a close.

In May 1867, Mr. Brassey, having occasion to go abroad on business,[1] was accompanied by a large party, consisting of the members of his own family. They first went to Paris. Interviews on business took up the greater part of his time; but all the intervals were employed in visiting the Exhibition, and in other sight-seeing, for he was as indefatigable in pleasure as in business, and was always most anxious that those who accompanied him should see everything that was worthy to be seen.

On the day that the party were to have left Paris, Mr. Brassey was very unwell, but was with difficulty persuaded to postpone his departure for a day. After travelling all night, and arriving at Cologne at four o'clock in the morning, he still insisted, though far from convalescent, that they should employ the time between four and nine o'clock in

---

[1] A description of these business tours by Mr. Tapp, Mr. Brassey's faithful and very intelligent secretary, is given in the Appendix. It is curious as showing the rapidity and energy with which their tours of inspection were conducted.

the morning, in seeing the Cathedral, and driving about the town.

Wherever he went, he was very well received; and when his arrival at any town was made known, many persons came to see him and to proffer civilities. In the remoter districts people would come many miles' distance just to shake hands with him, or to have a few minutes' conversation. When he came to any part of the country where the works of his own railways had been already commenced, his own people made every effort to show signs of welcome to their employer, sometimes clubbing together and producing a festive appearance with boughs of evergreens and festoons of coloured handkerchiefs. These primitive edifices bore mottoes, such as *Il Re degli Intraprenditori, Viva l'Imprenditore Inglese.*

Mr. Brassey made several trial trips on the engines of the Fell Railway, and his party went into the tunnel under the mountain; and when there, Mrs. Thomas Brassey could not help remarking "the large-hearted way in which he entered into its merits, the anxiety he expressed for its success, and the interest he took in its completion as a great enterprise and an extension of civilization." It is to be remembered that his wishes in this matter were directly opposed to his pecuniary interests, for the sooner the tunnel should be completed, the greater would be his loss on the line over the mountain.

Mr. Brassey returned to England on June 7; and, on October 7 in the same year, he started abroad again, though in a very feeble state of health, to attend the proposed opening of the Mont Cenis Railway, and also to transact some important business connected with the Lemberg and Czernowitz Railway. On October 10, he left Paris for St.-Michel, the weather being cold and tempestuous in the extreme—snow falling and a bitter north-east wind blowing. It was not, however, against bad weather only that Mr. Brassey, in an enfeebled state of health, had to contend. He had to experience ill-success, for the opening of the Fell Railway proved a most disastrous failure.

In the concession of this railway from the French Government a condition had been inserted, that the engines and carriages should be made in France. The engines

proved to be of the most faulty description. When the
first broke down, Mr. Brassey's disappointment and vexa-
tion were very great. He stood about in the cold and wet,
while another was telegraphed for. When it had arrived, and
had also failed, he was persuaded, though very reluctantly,
to return to the hotel at Lanslebourg, weary in body, and
depressed in mind, from perceiving the impossibility, for
the first time in his life, of keeping strictly to his word,
and opening a railway at the time when he had said it
should be opened. In the night he was taken alarmingly
ill with bronchitis; and in the morning was with great
difficulty carried on to Turin, in the hope of getting
medical advice there. Mr. Elliot and Mr. Tapp telegraphed
to Mr. Brassey's family, informing them of his illness.
They prepared to start at once; but hearing the next day
that he had gone on to Milan, and was proceeding to
Venice, they delayed their departure. They afterwards
found that the fever which had set in while he was at
Turin had affected his head, and in that state he was so
determined to go on that the doctors thought it better to
let him proceed than to irritate him by opposition.

When he arrived at Venice the fever was at its height.
His family received the most alarming telegrams and immediately started, scarcely having a hope of seeing him
alive. They never stopped *en route*, finding a special train
at Macon, and post-horses ready for them at every stage
over the mountain (Mont Cenis), and everybody anxious
and striving to do their best to expedite their journey to
the utmost, that there might be no delay. When they did
arrive, however, though finding him still in a critical state,
there was every prospect of his recovery. After that he
rallied in a marvellous way, and while still in bed contrived
to transact his Lemberg and Czernowitz business.

About the second week in November his family were
able to bring him home by slow stages to St. Leonards,
where he remained quietly for ten weeks to the great benefit
of his health, transacting his business there, without going
to London.

This, however, was the beginning of the illness which
eventually proved fatal to him.

In 1868 he had a second stroke of paralysis. This,

however, did not prevent him from continuing his labours with the same assiduity as ever. When urged to restrict those labours, he would not do so, feeling that any remissness or relaxation on his part might occasion a loss of employment to those who had for so many years so faithfully and so zealously been fellow-workers with him.

It is a curious circumstance that the last contract work which Mr. Brassey undertook, and which he himself personally supervised, brought him close to the spot where he had commenced his railway undertakings.

His first railway contract was the Penkridge Viaduct. Among his latest was the Wolverhampton and Walsall Railway; and probably the last work he ever did was to survey some construction a few miles distant from the Pankridge Viaduct.

Some time before his death Mr. Brassey knew that his disease was fatal; and he bore that knowledge with resignation. He had ever been a very religious man. His religion was of that kind which most of us would desire for ourselves,—utterly undisturbed by doubts of any sort, entirely tolerant, not built upon small or even upon great differences of belief. He clung resolutely and with entire hopefulness to that creed and abode by that form of worship in which he had been brought up as a child.

The tender-heartedness I have mentioned as a pervading element in Mr. Brassey's character, had never been more manifested than in his conduct on the occasion of any illness of his friends and associates. At the busiest period of his life he would travel hundreds of miles in order to be at the bedside of a sick or a dying friend, and to give what aid or consolation he could give to him. It was now his turn to experience similar kindness; and never was illness watched with more solicitude and affection than was Mr. Brassey's last illness.

A touching proof of the regard and affection, which Mr. Brassey had won from those who had served under him, was manifested during this illness. Many of these persons, both those who had served him in foreign countries and at home, came from great distances, solely for the chance of seeing, once more, their old master whom they loved so much. They were men of all classes;

humble navvies as well as trusted agents. They would not intrude upon his illness, but would solicit to be allowed to stay in the hall, and would wait for hours there, in the hope of seeing Mr. Brassey borne to his carriage, and getting once more from him a shake of the hand or the slightest sign of friendly recognition. The world is, after all, not so ungrateful as it is sometimes supposed to be; those who deserve to be loved generally are loved, having elicited the faculty of loving which exists, to a great extent, in all of us.

The air of Hastings was favourable to the invalid; and it was there that he spent the last days of his life. His disease, however, was not to be overcome by change of air. He grew rapidly worse; and, after much suffering, expired, surrounded by his family, on December 8, 1870. Mr. Brassey was buried in the churchyard, at Catsfield, Sussex. He left behind him as members of his family, ever to deplore his loss, his widow and three sons, Thomas, the Member for Hastings, Henry Arthur, the Member for Sandwich, and Albert.

On Mr. Brassey's death the grief of his friends was great and unanimous. They felt that in losing him, they had lost one who gave a hearty welcome to them, whether they came to impart their sorrows or their joys, and who was equally ready in either case to give them aid, counsel, encouragement, and sympathy.

CHAPTER XXIII.

RAILWAYS AND GOVERNMENT CONTROL.

THE subject of Railways is not by any means exhausted. The world will yet have a great deal to think and say about them; and this, to my mind, gives additional interest to the story of Mr. Brassey's labours.

It were much to be wished that the intelligence and skill shown by the contractors for the making of railways, and notably by Mr. Brassey as the most eminent of them, could have been shown in the management of railways after they had been made. I fear, indeed, we must confess that, neither in the promotion of railway schemes, nor in the investigation of these schemes by Parliamentary Committees, nor in the subsequent administration of railway property, has anything like the skill and intelligence which were manifested in their construction been exhibited. In fact, there has been a deplorable want of organization in all railway affairs, with the sole exception of the skill exhibited in their construction.

In Great Britain the promotion of schemes has, in many instances, been entirely devoid of any of that consideration for the public welfare which such a matter imperiously demands. Then, as to the expenses attendant upon getting a Railway Bill through Parliament, these have borne hardly any proportion to the inherent merits or demerits of the schemes themselves. The legal expenses incurred during the height of the railway mania were a disgrace to civilization. It was never more clearly shown how inadequate Parliamentary Government is to deal with questions of administration, when these come before it suddenly and without sufficient previous consideration by those thinkers and writers who really direct the public mind. At the time of the railway mania a state of things prevailed which very closely resembled that produced by the celebrated transactions of John Law; and I do not think that it was dealt with more wisely than Law's speculations were dealt with by the uninstructed ministers of the Regent of France.

If we turn to what has been done since in the administration of railways, we shall often find that the public convenience has not only been neglected, but has been industriously and carefully thwarted. The times for the arrival of trains have been so arranged as *not* to correspond with the times of starting of the trains on other railways, which, for the public convenience, ought to have been in close correspondence.

The remedy for this evil has only produced a new disease. This remedy has been the amalgamation of railways; and though certain injurious effects of competition have thereby been removed, that end has been attained by the adoption of a system which takes little heed of the interests of the public. In short, these injurious effects of competition have been occasionally removed by the creation of monopoly.

I may here mention, as a most important fact, that Mr. Brassey thought that the admission of this principle was an error in judgment. He was of opinion that the French policy, which did not admit the principle of free competition, was not only more calculated to secure the interests of the shareholders, but that it was more favourable to the public. He, moreover, considered that a multiplicity of parallel lines of communication between the same termini, and the uncontrolled competition in regard to the service of trains, such as exists in England, did not secure so efficient a service for the public as the system adopted in France. Mr. Thomas Brassey says that he remembers, when travelling in France with his father, that Mr. Brassey constantly lemarked the superior comfort of the carriages, the excelrence of the stations, and, generally, the superior amenities of railway travelling abroad as compared with those in England. He would then express his regret that the policy, which had proved in detail so successful in France, had not been adopted in England. He thought that all the advantage of cheap service, and of sufficiently frequent communication, which were intended to be secured for the

British public under a system of free competition, could have been equally well secured by adopting the foreign system, and giving a monopoly of the network of railway communication, in a given district, to one company; and then limiting the exercise of that monopoly by watchful supervision on the part of the State in the interest of the public. With regard to extensions, Mr. Brassey held that the interests of the public would be perfectly secured under the French system; because it was quite possible, as has been actually proved by the extension of the French railway system, to secure, on the part of the Government, an authority which would enable them to compel the Railway Companies to extend their system gradually over the whole districts in which they had a monopoly of the railway communication. This process, having been carried on gradually, has been found in France to be compatible with the maintenance of a fair average rate of dividend to the original shareholders in railway enterprises.

I have given these opinions of Mr. Brassey in full, but I am not prepared to support them; and, indeed, I have not the requisite knowledge or experience to pronounce any decided opinion upon so difficult a subject. If I were obliged to give an opinion, I should say that there had, undoubtedly, been a considerable dereliction of duty on the part of our Government and Parliament as regards the control, which they failed to exercise, both for the benefit of the public and of the shareholders, if, indeed, these interests are distinct. On the introduction of the railway system into Great Britain, I think that this control might have been inaugurated without any serious or lasting interference with the principle of free competition.

One most important statement I must adduce here as regards the conduct of the late Sir Robert Peel, in reference to this matter. It is a statement that was made by himself. He said that he had not at his command sufficient power (he meant official power) to institute a control over these undertakings. I believe that what he said was exactly true; and it has always appeared to me to be one of the most memorable instances of the hide-bound nature of our official system, which is hampered by so many checks and so much dread of small expense, that the most needful undertakings have to be passed by, or touched but lightly, which require the best intellectual force of the nation to be brought to bear upon them.

The amount of money expended in Great Britain on railways is nearly £500,000,000, and it would have required about 1*d*. per cent. to have gained the official power requisite for efficient control of those railways in the interests of the public. I put aside for the moment the immense injury done to private individuals by the limitless extension of railway works with unnecessary and thoughtless rapidity.

The questions concerning railway management will gradually force themselves upon the consideration of the public; and it is noticeable that already many thoughtful persons have come to the conclusion that all railways should belong to the State, and be under one central control. This, however, is a very serious conclusion; for, unless Government is stronger than it appears likely to be in our time, it will hardly have power to make head against the criticism and the odium to which it will be subjected immediately upon its having the management of such a vast and complicated concern as the railway traffic of this country. There is not any subject of social interest which requires more thought than this question of absorbing into the functions of Government the whole of our railway system. Whatever has been done in a similar direction— as, for instance, the transmission of letters by the Post Office, and, in our time, the taking over of the Electric Telegraph systems by Government—are comparatively small matters when put side by side with the question of the Government undertaking to manage all the railway traffic of the country.

I am not unaware that those persons who defend the course which has been taken in England as regards the promotion of railways, have a great deal to say for themselves. They contend, and justly, that if the whole matter had not been thrown open to competition, many lines which are now useful, and which have promoted commerce and developed industry, would never have been made. But I think that even these persons would admit that, having had the benefit which this unlimited competition has produced, it is now a subject open to good argument, whether a consolidation of the whole system might not take place under the direct management of the State.

When contemplating the errors which most persons will allow have been made in the primary promotion of railways, in the conduct of railway Bills through Parliament, and in the management of railways after they were made; and, on the other hand, when considering the skill that has been used, and the resources that have been brought to bear, in the mere construction of railways; a simile will naturally occur to the mind, of this description. You see a child, a little thing perhaps of six years old, leading and guiding at its will a team of magnificent horses. This sight forces upon the mind a feeling of satisfaction at the immense preponderance of the human intellectual power, as shown in this child, over the brute force of these powerful creatures. Such has been the sway of capital, often very ignorantly, childishly we may say, applied in ordering the construction of railways. Schemes recklessly promoted, or at any rate promoted without any view to the general good, have been efficiently carried out by the docile contractor—whom, by the way, I am rather ashamed to compare for the moment, except for faithfulness and docility, to the horse. But it is now time to consider whether the grown-up man should not take the place of the child, and assume the guidance of all that work, misapplied sometimes, but nearly always well accomplished, which has now been brought, completed, to his hands.

There is a special subject connected with the establishment of railways in Europe, which will require very careful consideration, and, at some time or other, very prompt action. As we all know, there is no shadowless good in any of man's inventions. Undoubtedly great advantages have accrued to the world from the introduction of railways. These rapid means of transit have not only promoted commerce and enabled human beings to communicate more freely with one another. They have also caused diseases, both of man and beast, to be propagated more freely. In fact, this swift mode of communication may quietly effect, in relation to the propagation of disease, that which was tumultuously effected in former times by the movement of great armies. This condition of things has recently been made manifest by the Spreading of the Cattle Plague. There is no doubt that the facile mode of transit caused by railways, places us in constant danger of the introduction of infected Steppe cattle, or of others which have taken the infection from those Steppe cattle. How successful this dreadful disease has been in forming for itself a home in Russia may be seen from the incantations, the charms, and the songs which have long existed in Russia in reference to cattle plagues, and which have lately been made known to English readers by a work wherein special mention is made of this branch of Russian literature.[1]

But this is not all. It is justly feared that some of the diseases most fatal to man will find, and even now are beginning to find, an easy access from their haunts in Eastern countries to the hitherto more favoured nations of the North and West. This danger is attracting the attention of some of the most thoughtful persons, holding sanitary offices under some of the principal European Governments; and it is a matter respecting which those Governments will probably be obliged to take action.

The writer may perhaps be permitted to observe that the rise of such a question tends to illustrate the proposition which he has maintained elsewhere; namely, that increasing civilization continually produces more work for Government. For instance, it will not be contended that quarantine regulations are matters which private individuals can provide for themselves. I therefore deem it to be unquestionably desirable, in a work which has to deal so largely with questions concerning the introduction of railways, to point out any difficulty or danger which has been

[1] "The oldest woman among them is then yoked to a plough, and she must draw it three times round the whole of the village, the rest of the party following after her, and singing the songs set apart for such occasions. It is supposed that the malignant spirit, whom they recognise in the cattle plague, will be unable to cross the lines thus traced by the plough, or to get at the cattle, which during the ceremony have been kept shut up within the village."—W. R. S. RALSTON, *Songs of the Russian People*, London, 1872.

caused by that introduction, in itself so great a proof of human skill, and, upon the whole, so favourable to the happiness of mankind.

The remarks in this chapter relating to the past inaction of Government as regards railways, have not been made in a cavilling spirit, or as a mere ungracious expression of regret that more foresight was not manifested when railways were beginning to be made in this country. To have laid down, in the first instance, a general plan for the introduction of the railway system; to have provided for its judicious extension; to have contrived less costly means, and a less uncertain tribunal, for investigating the relative merits of proposed railways; also to have constituted some central authority, endowed with sufficient power and sufficient knowledge to protect the interests of the public,—would not have been objects unworthy of the attention of Government, and would have exhibited some of that foresight which is lamentably deficient, for the most part, in nearly all governments, even in those of the most civilized nations. It now only remains to furnish remedies for the evils in the system which have arisen from its having been entirely abandoned to private enterprise, evils which have marred, to some extent, the advantage arising from a series of the most splendid works of construction, exceeding in magnitude and utility all the public works that had hitherto been made upon this earth, whether by despotic monarchs with their millions of slaves, or by the united labour and intelligence of highly civilized communities.

APPENDIX A.

MR. TAPP'S NOTES ON MR. BRASSEY'S TOURS.

DURING the last few years of Mr. Brassey's life it was my duty to accompany him on his visits to Austria, made generally in the months of May and October, to inspect the works of the various lines in which he was interested.

We usually went by way of Paris and Strasbourg, and returned *viâ* Venice and Turin, and so over the Mont Cenis, where the works of the Fell Railway were in progress, or we went by Mont Cenis, and returned *viâ* Bavaria and Strasbourg.

On one occasion Mr. Brassey's partner, M. Schwarz, and his agent, M. Fölsch, met him by appointment in Vienna, and a day or two after left with him to go over the works of the Crown Prince Rudolf Railway in Styria. As the firm, Brassey, Klein Bros. and Schwarz, were engaged in the execution of a number of lines running up the different valleys in Styria, some of which were nearly finished, some just commenced, one just opened for traffic, and others only contracted for a few weeks before, it was necessary to go over the greater part of Styria in the course of our journey, sometimes in carriages, at others in a railway carriage, and for not a little of the distance in an open ballast wagon, fitted up with temporary seats, and this conveyance seemed to suit Mr. Brassey best, as he could see all the line from one end to the other.

On leaving Styria we went over the Predil Mountain, through which a long and difficult tunnel was projected, which Mr. Brassey said would be as difficult of execution as the Mont Cenis. On going over the mountain, we descended to Gortz on the shores of the Adriatic; and the sudden change, from the cold barrenness of the country on the north side to the warm climate of the south, the villages interspersed with orchards, the trees in full bloom, was very striking. Mr. Brassey, who left London in very indifferent health, seemed to feel the influence of the mountain air in a remarkable degree, and said we had chosen the very best week of the year for our visit. At all the places where we stopped to bait the horses or to sleep, a repast on a grand scale had been prepared by the agent of the district, under M. Schwarz's directions, and the resources of the neighbourhood taxed to the utmost to do Mr. Brassey honour. At one place, the largest trout that had been caught for many years was served up for supper, *cold*. We stayed a day at Trieste, to answer the letters which had been accumulating there from London for some days, and then returned to Vienna by way of the Soemmering. On the afternoon of our arrival we left for Pesth, where Mr. Brassey had to meet Sir Morton Peto; and, after spending a very few hours, we then went on board the steamer to go down the Danube as far as Giurgevo. The river journey occupied about three days and three nights, including stoppages.

At Giurgevo, Isidore, Mr. Brassey's courier, hired a vehicle, half cart half phaeton, covered with a long leather head, which sheltered passengers, luggage, and driver, driven by four ponies abreast, to take us to Bucharest, about forty miles distant. The greater part of this journey was done at a gallop, and I believe it occupied five hours, including an hour to bait the ponies. Mr. Brassey was delighted with the coachman and the ponies, saying it equalled the best days of English stage coach travelling, and told Isidore to buy the carriage and ponies to use for the rest of the journey, which, however, he was unable to do, as they had to return to Giurgevo.

Arrived at Bucharest Mr. Brassey found M. Ofenheim and Mr. Strapp, his agent, who had come over from Austria to meet him, and they remained a week in the capital of the Principalities, to transact some business with the government of Prince Charles. At the end of that time we left for Jassy; and as our party had by this time considerably increased in number, we started in two carriages, one drawn by eight, the other by six horses, and drove for six or seven consecutive days through Wallachia and Moldavia, staying at night at the houses of the landed gentry or Boyards, where we invariably met with a most hospitable reception. Some part of this journey was done at the rate of twelve and a half miles an hour. I timed the horses very carefully and found we were exactly three minutes doing the distance from one kilomètre post to another.

At Jassy, the ancient capital of Moldavia, Mr. Edwards had an office, and there was a staff of English, so we felt, as it were, in our own country. A grand reception was prepared for Mr. Brassey here, flags flying, some very small cannons firing, and the priest of the village, and a number of workmen with their families, in readiness to receive him. From this place to Czernowitz Mr. Brassey was accompanied by M. Ofenheim, his agent, Mr. Strapp, and others, in an examination of other parts of the line; and after staying a few days at Czernowitz, we left for Lemberg and Cracow. Here we had the use of M. Ofenheim's saloon carriage, the most convenient conveyance for railway travelling I ever saw.

It contained a stove, two sofas, reversible at pleasure so as to become beds; a most comfortable writing desk, with every appliance for conducting correspondence, while commanding a complete view of the line, permanent way, stations, &c.; a table, a cellar of wine packed away in ice; an aneroid barometer to indicate the changes of gradient; a dressing-room, furnished with every convenience for the toilet, even to a shower-bath, and arm-chairs, allowing the occupants of the carriage to change the position of their seats; lastly, a complete section of the whole line was pasted on the sides of the carriage, showing every bridge, culvert, or level crossing, &c. In this carriage we travelled three or four hundred miles without fatigue.

In these journies nothing struck me more than Mr. Brassey's great economy of time. He generally advised his friends by telegram of the day of his arrival, so that some business was ready to be transacted the next day. His mind seemed always employed; and as the May journey generally took place just after the balance sheet for the previous Christmas had been made up, I had to answer from memory a great many questions connected with his various contracts, his commitments, the liabilities on each of them, and the assets available, or otherwise. He had a remarkable way of carrying all these figures in his head, and always knew the probable result of each undertaking. Nothing seemed to escape him in the country we passed through; the systems of cultivation, &c., always seemed to interest him.

He was well received everywhere; and it often struck me that his name was a kind of passport which opened all doors.

When Mr. Brassey was staying at Paris or Vienna, he gave

frequent dinner parties at hotels where he stayed, inviting such of his staff as were living near, and on some occasions the directors of the lines he was engaged upon, bankers and others with whom he had business relations. At these meetings he was most genial, and the hotel-keepers did their utmost to please him, and looked for his arrival from London with much apparent satisfaction.

The journey alluded to above occupied seven and a-half weeks, April 17 to June 5, 1869; and that undertaken in October of the same year was the last Mr. Brassey made to the Continent. Although he was at the time very far from well, never having recovered thoroughly from the illness he had at Venice, the journey was a very rapid one, rarely staying more than one day in a place. We must have travelled over between five and six thousand miles of country from the time we left till our return.

APPENDIX B.

LETTERS.

TO MR. GOODFELLOW.

May 8, 1851.

DEAR GOODFELLOW,—I wish you to read carefully the enclosed statement relative to the accident that happened to ——. I think myself it is quite immaterial whether the piece of iron was thrown down or fell down by accident, as affects our liability. If the iron caused the damage, and it fell from our scaffolding on the public highway, I take it we are liable; and I should recommend the best settlement to be made that can be without reference to lawyers.

Will you, after reading the statements I have enclosed, let me have your opinion upon the matter, or, what would be far better, settle the affair if you can do it reasonably.

* * * * * *

I am, my dear Sir, yours very truly,
THOMAS BRASSEY.

MR. GOODFELLOW.

TO MR. DAY.

January 27, 1859.

MY DEAR SIR,—R—— M——, who you know perfectly well, has been here repeatedly enquiring for work, and is very desirous to get to you on the Severn Valley.

If you have no objections to him, write and say that when you have more land an opportunity shall be given to him to give a tender. But if you do object, then write to say so, for at present he is waiting under the impression that he will be told as soon as any more land is obtained.

He knows his work well, but by some means he got into a suit or reference against me on the Leicester and Hitchin, brought on entirely by continued payments on account, instead of monthly settlements, and the death of poor Mr. Hone.

I am, my dear Sir, yours, very truly,
Thomas Brassey.

Mr. Day.

TO MR. DARKE.

July 27, 1864.

My dear Sir,—The prospect of the cotton trade appears very encouraging, for some time to come at all events, and very tempting. But I have carefully reflected upon the business, and have come to the conclusion that if successful I might be tempted to go on to a large extent, and should in fact become a large cotton speculator, which I have no desire to become. If, on the other hand, I was to make a loss, which is quite possible, I should feel annoyed that I had departed from my legitimate business, and I therefore come to the conclusion not to enter into the speculation.

Thanking you much for calling my attention to the subject,

Believe me, my dear Sir, yours very truly,
Thomas Brassey.

Thomas E. Darke, Esq.

TO MR. HOLME.

May 25, 1865.

My dear Friend,—I returned yesterday after three weeks' absence in Italy (where I have been to the inauguration of the Maremma Railway), and found your kind letter enclosing Mr. ——'s, which I return.

I have much pleasure in enclosing a check for £100, which Mr. —— will dispose of as he pleases.

I am glad to know that you are getting established at Southport, which from all that I have heard is, for the north, extremely mild, and on that account will, I trust, suit Mrs. Holme and yourself, for at all events the greater part of the year. It must, I know, cost you a very hard struggle to give up your public functions which you have so very ably and honourably filled for a great number of years; but we have all our day, and time is very short. For myself; I appear to be driven along with the tide of affairs from which I find it difficult to extricate myself, you will say, and very justly, from want of moral courage. Such I feel it to be, and I do promise henceforth to be more firm, which promise I hope to perform.

The weather in Italy was really magnificent, for a week at least, and here everything looks very fresh and beautiful.

With my very kindest regards to Mrs. Holme and your daughters,

Yours very truly,
THOMAS BRASSEY.

SAMUEL HOLME, ESQ.

TO MR. MURTON.

56 Lowndes Square : June 29, 1868.

MY DEAR MURTON,—I am sorry I have delayed replying to your favour of the 27th inst. so long, but on Saturday I had a second attack, which prevented my attending to it, and your note was laid aside, and only turned up this morning.

I do not know whether Mr. —— could find Mr. —— anything. Will you call in Great George Street, and see him if he has not returned. If he has, write to him. I should be glad to do what I could for Mr. ——, but I really do not see my way unless —— could find him something temporarily.

Tell —— to prepare a check for twenty pounds for my signature; it will relieve his immediate necessities.

I, as you will see, am suffering a good deal from my attack, in the free use of my hands; still I hope you will make out what I have written.

Yours very truly,
THOMAS BRASSEY.

[The delay alluded to was only that of from the 27th to 29th, the 28th being Sunday; and Mr. Brassey was confined to his room by a second paralytic attack. He notwithstanding took the trouble to write the above with his own hand, and evidently with much difficulty.]

FROM MR. MILROY

8 Salisbury Road, Edinburgh : October 10, 1871.

MY DEAR SIR,—I have much pleasure in sending you the following brief notice of the late Mr. Brassey, as he appeared to me through the long period of more than thirty years during which I had the honour of his acquaintance.

His transcendent ability is too well known, and finds too notable a monument in his great success, to require much illustrative comment. From first to last it was calm and modest: the most complicated affairs were conducted with so much ease, and were disposed of so quietly and yet so completely, as to convey the impression that their difficulty was not so great as it appeared. This ease was due not only to the rapidity and unerring accuracy of his judgment, but also to the possession of that characteristic of all great minds—the faculty of sufficiently inspecting details, and at the same time of taking a wide and comprehensive view of the whole matter under consideration. In negotiation he also had the advantage of diplomatic ability of a high order. His superiority in this respect had nothing in it of the arts of the trickster: it lay in the natural influence of a man of consummate talent, shrewdness, tact, urbanity, and straightforward integrity. At a conference where conflicting interests were represented, I have seen this power markedly displayed. In the course of the conversation some one hinted at an evasion of his obligations on the ground that a legal formality had been neglected in the arrangement of an agreement. To have discussed or disputed the quibble would certainly have led to a long argument, and might have frustrated the object of the meeting; but it was at once met by a remark from Mr. Brassey to the effect that amongst honourable men such a view of the case was out of the question—the remark being made in such a way as to make everyone, the objector included, feel that nothing more need be said.

His scrupulous and honourable discharge of every obligation resting upon him was always to my mind a very prominent characteristic—one which sometimes seemed even to be carried to an extreme. I need scarcely say that the best of materials and the most skilful workmanship were invariably procured; and, though every arrangement or contrivance calculated to facilitate or cheapen construction met with his pleased approval, anything like bad or dishonest work was resolutely discountenanced. Instances, again, which came within my own knowledge, might be given of claims which might have been made on railway companies for delay on their part, and for the expense which he had in consequence incurred; but, though brought under his notice, these claims were never sent in. To urge such demands " was not his way of doing business." The fall of the Barentin Viaduct on the Rouen and Havre Railway brings out the same generosity of feeling. In 1846 the original structure, of light and elegant design, had nearly reached completion, when one night it fell to the ground from end to end, owing to some defect, which he himself had previously pointed out in an offer to share with the Company the expense of the necessary remedy. This catastrophe involved a loss of at least £30,000, which in those days was not a small sum even to Mr. Brassey; and as there were good reasons for laying the blame and the loss on others, most men would in the circumstances at least have disputed their liability. Not so Mr. Brassey. Without pausing for a moment on such questions, he at once addressed himself to the task of filling up the gap in the railway communications which the fall of the viaduct had caused, ordered the *débris* to be removed, and arrangements to be made for the reconstruction of the work according to the new design which the engineers had prepared. The work was pressed on with so much energy that the ruins were removed, an entirely new viaduct was built, and the trains were running over it, within six months from the commencement of the operations. A recent French guide-book to the Western Railways of France speaks of this viaduct having been rebuilt " avec une rapidité merveilleuse."

While thus chivalrously exacting in the discharge of his own obligations, he was not, as is too often the case with such a character, severe in dealing with other men. For instance, early in his career, he had let the construction of a wooden bridge over a river to a sub-contractor for a " lump sum." The latter, on proceeding to put in the foundations, found that cofferdams and pumping of an expensive nature were indispensable, and for these unfortunately he had made no sufficient provision in his estimate. Alarmed by the prospect of losing seriously by the contract, he wrote to Mr. Brassey, frankly stating his difficulty. Somewhat to his surprise, and much to his relief, Mr. Brassey's prompt answer, based probably on a conviction of the truth and honesty of the appeal, was that he would take upon himself the expense of the cofferdams, &c. and that the work might be continued on that understanding. This anecdote is quite in keeping with his usual way of dealing with those whom he employed. In letting various bridges and cuttings on a line in France I suggested that the agreements made with the different sub-contractors should be embodied in binding documents, according to the forms prescribed by the French law. "Well," said Mr. Brassey, "if it is clearly understood what is to be paid for the work, I think it is scarcely necessary. For if a man is competent and has a fair price, he will make the job pay, and accordingly will not try to get rid of it; if he has not an adequate price, the sooner we make it right for him the better. On the other hand, if he is either idle, incompetent, or troublesome, the sooner we get rid of him the better." The policy thus sketched out was dictated, not merely by natural wisdom, but also, I believe, by a goodness of heart which rose frequently to magnanimity. The following illustration will show this. A sub-contractor, R. M., as well as other members of the same family, had reccived numerous benefits from Mr. and Mrs. Brassey during a period of many years. R. M. latterly, however, became utterly unreasonable, having made some friends whose advice was more ingenious and grasping than sound and becoming. Under their influence, on the completion of several small contracts, he repeatedly made extravagant claims for more money, on many pretences for which there was no foundation except in his own imagination. It was of course refused, and R. M. was foolish enough to resort to legal proceedings. Through every avenue of litigation he pestered Mr. Brassey for a long time, until at last, after having been beaten at every previous stage, his absurd claims were effectually set at rest by a final decision. Notwithstanding the trouble and annoyance he had undergone, Mr. Brassey, on the day when the cause was decided, said to me, "I am afraid, Milroy, that after this litigation, R. M. must be badly off; I wish you would get him a job. I would be glad if you could."

As might have been expected of such a man, Mr. Brassey was gifted with an eminently cheerful disposition, which enabled him to be happy in any circumstances. I remember that he came down to Normandy the day after he had dined at the Tuileries with the Emperor Napoleon, to visit a tunnel the construction of which was attended with some difficulties. In order to explore the workings, which were dripping with water, he donned a rough miner's dress and cap, which together formed a costume the reverse of imposing and elegant. After the exploration of the tunnel, when everyone was tired, wet and hungry, the only refreshment which could be procured was bread and cheese, which were discussed by the company in a shed, in ploughman fashion, with the help of their pocket-knives. No one laughed more heartily than Mr. Brassey at the amusing features of the situation, especially when he contrasted his own costume and the fare with the studied elegance, the gold plate and rich dishes, of the Imperial banquet a day or two before.

Before closing this letter, I will only add that all connected with Mr. Brassey could congratulate themselves on his urbanity and kindness. Nothing but a natural courtesy which had settled down into a habit, and was aided by an excellent temper, could have withstood the constant worries of so busy a life. None of his correspondents, agents, or friends, ever applied to him for instructions or advice, without an instant response on his part: whether he were in Russia, Turkey, or Spain, his industry and loyalty were such that a reply was at once despatched. Altogether he was a man for whom one could hardly fail to conceive a deep feeling of respect and esteem.

I beg to take this opportunity of recording my grateful remembrance of his kindness to myself, and remain, my dear Sir, yours very faithfully,

JOHN MILROY.

T. BRASSEY, ESQ., M.P.

FROM MR. HAWKSHAW.

33 Great George Street: December 11, 1871.

MY DEAR SIR,—I fear you will think I have forgotten my promise to write you a few lines on the subject we spoke about in reference to the career of the late Mr. Brassey, viz., the origin of our large contractors, and the relative value of the unskilled labour of different countries.

With regard to large contractors, they were not unknown before the railway era. The construction of inland navigation, and of docks and harbours, had called them into existence to a limited extent; and Messrs. Joliffe and Banks, for instance, were large men of that class.

But with the commencement of the railway system began an age of great works, during which undertakings of far more colossal dimensions were rapidly projected, and required to be as rapidly carried into execution. The extension of the railway system called also for larger docks and larger harbours, and since the construction of the Liverpool and Manchester Railway the public works of all kinds that have been executed in the United Kingdom alone far exceed all that had been done before.

But though railways originated in England, they rapidly extended to other countries; and in the first instance many of the continental lines were intrusted to the skill of English engineers, and were constructed by English contractors. Thus within a comparatively short space of time—less than one generation—there arose a demand for men of constructive skill and ability such as had never been known before, and if the public works since then be compared with the public works of all preceding time, it will show that the latter are comparatively insignificant.

And Mr. Brassey's labours were not confined to this country alone, but, like those of some others of our large contractors, extended to nearly every quarter of the globe.

In this, as in other cases, the occasion created the men, as great occasions always will do.

Men were wanted to design these works and to execute them, and engineers and contractors sprang, as it were, from the earth on whose surface they were going to make so much impression. There was no time for preliminary education. The demand arose, and men who felt or thought themselves capable to undertake the duties stepped forward from every rank. It was by a process of what Mr. Darwin has termed "natural selection" more than in any other way that they were found. They felt themselves in that line to be the strongest; and how true were their instincts, no one is a better illustration than our lamented friend.

On the other subject, viz. the relative value of unskilled labour in different countries, I have arrived at the conclusion that its cost is much the same in all. I have had personal experience in South America, in Russia, and in Holland, as well as in my own country; and as consulting engineer to some of the Indian and other foreign railways, I am pretty well acquainted with the value of Hindoo and other labour; and though an English labourer will do a larger amount of work than a creole or Hindoo, yet you have to pay them proportionately higher wages. Dutch labourers are, I think, as good as English, or nearly so; and Russian workmen are docile and easily taught, and readily adopt every method shown to them to be better than their own.

With regard to unskilled labour men seem to be like machines : the work given out bears some relation to the food consumed. A good illustration of this occurred on the French railways executed by Mr. Brassey.

When he commenced the Paris and Rouen Railway, he began by largely employing English navvies, paying them much higher wages than would have been required by French labourers, but the larger amount of work done by the Englishmen compensated him for the higher wages.

After a time, as I have heard Mr. Brassey say, the Frenchmen, gradually receiving higher wages than previously they were accustomed to receive, were enabled to live better and to do more work, until ultimately the French labourers came to be chiefly employed.

In conclusion, I have only to add a few words on Mr. Brassey's personal qualities.

He possessed great powers of mind, with an almost intuitive perception (when dealing with large transactions) of final results. He had withal a most kindly disposition.

Though he amassed a very large fortune, it was not in undue proportion to the vast magnitude of the many undertakings he was engaged upon; and he was proverbially liberal to his agents, and those in his employment.

While he acquired great wealth, he was ever wholly free from ostentation, and never lost his original simplicity and modesty of character; and he left behind him a name untarnished by aught unworthy of an honourable man.

I am, dear Sir, yours truly,
JOHN HAWKSHAW.

ARTHUR HELPS, ESQ.

FROM MR. HOLME.

23 Royal Crescent, Bath : March 18, 1872.

MY DEAR MR. WAGSTAFF,—I must sit up to write you a line.

It has been said that biographers always present the fair side only of the characters they delineate, and there is much truth in the observation : but with Mr. Brassey the observation is negative;

for in all his transactions with his fellow men, it would be impossible to find a dark shade as a foil to the brightness. During our forty-six years' friendship, in which I saw him under all circumstances, I never discovered the slightest tinge of error, nor the slightest approach to selfishness, meanness, or love of applause. I have never heard him say an unkind thing of any human being, but I have known him make sacrifices to oblige and benefit others, which betokened the most generous feelings.

I never mentioned a case of need to him without receiving prompt and liberal aid, and a member of my family, who highly appreciated his generosity and forbearance, used to say, that Mr. Brassey's favour might be obtained by doing him an injury, only that no one could be found who would or could so act after knowing him.

I have had much knowledge of, and intercourse with, mankind of all classes: but our friend stood out so prominently for simplicity and wisdom, energy and self-abnegation, prudence and generosity, sincerity, largeness of mind and honour, that I always considered his example to be valuable, and his personal friendship a blessing.

I have written this with much difficulty, and must now get on my sofa.

With much regard, believe me, my dear Sir,
       Yours very sincerely,
         SAMUEL HOLME.
W. WAGSTAFF, ESQ.

FROM MR. WHEELWRIGHT.

Gloucester Square, Regent's Park: April 4, 1864.

MY DEAR SIR,—When you said to me on the day you made the arrangement with the Financial Companies, that you had done it on my account, in consideration of the long years I had laboured to accomplish it, I did not know how to thank you or how to express my gratitude to you, for I was sure you expressed the honest sentiments of your heart, and I hope and trust that I shall always recognize your kindness. Nothing would grieve me more than to find you inconvenienced from doing such an act of disinterestedness.

It must and will be satisfactory to a benevolent mind like

yours to know that you are conferring upon that vast and prosperous country such an infinite benefit, for to you will belong the credit of carrying out this great undertaking, the greatest by far that has ever been initiated in all the South American Continent, for it is but the beginning of a line that will penetrate into Upper Peru, and finally reach the Pacific Ocean, and bring together nations of the same race and blood, that have for centuries been isolated from each other by the Cordilleras of the Andes, and as Admiral FitzRoy declared before the Royal Geographical Society, it will form the shortest and best route between Great Britain and Australasia. Indeed, when I reflect on the incalculable benefits it will confer on the world, I am quite lost in the broad spread field it will open up for the trade and commerce, as well as for the civilization and well-being of millions of people.

Looking at it in a pecuniary point of view, I have studied the matter for more than ten years, and so far from having been able to discover any flaw in it, I have been more and more convinced of the soundness of the scheme and its legitimate value, and have not been alone in the estimate of this value. Many sound and judicious men have investigated the subject, and I have found but one opinion as to its merits. We go into the heart of a country abounding in mineral, pastoral, and agricultural wealth, possessing a soil and climate unsurpassed, and capable of producing all the wants of man. Some there are who will urge the political insecurity of the country, but the only answer I can give to them is the fact, that, in spite of the dissensions and revolutions, the country has gone ahead in a most rapid manner. The products and consumption of the country have immensely increased and continue to increase more and more rapidly.

The European element of population which is very great now will in a few years exceed the native; these are the best tests, as regards its security, for who would go if property were insecure?

The yearly emigration is augmenting in an extraordinary ratio; and if all that is inviting in that favoured land were known the increase would be much greater.

As to the cost of the road, I know enough of railways in South America to give an opinion founded upon experience; and if we do our work as we should, that is with a close attention to economy, you will find all the calculations will be fully and completely realized.

I have also given you an honest opinion as to the great value of the land; and if we can but manage it properly, and not be in too great a hurry to sell, except a small part for the encouragement of settlers, much more will be realized than we contemplate.

Our titles are clear and investigationable. We have at least the halo of British protection, which is the only true way of viewing it; and renewing my warmest thanks for what you have done,

    Believe me,
      Truly and faithfully yours,
          W. WHEELWRIGHT.

THOS. BRASSEY, ESQ.

APPENDIX C.

TABLE

SHOWING THE WAGES PAID, FOR RAILWAY AND OTHER CONSTRUCTIONS, IN VARIOUS PARTS OF THE GLOBE.

England.

	Per diem.			
	s.	d.	s.	d.
Gloucester and Bristol Railway (1843):				
Labourers	2	4 to	2	9
Great Northern Railway (1849):				
Labourers	2	9 ,,	3	0
Shrewsbury and Hereford Railway (1851):				
Labourers	2	4 ,,	2	9
Leicester and Hitchin Railway (1855):				
Labourers	2	9 ,,	3	3
Shrewsbury and Crewe Railway (1858):				
Labourers	2	9 ,,	3	0
Shrewsbury and Hereford Railway (1861):				
Labourers	2	9 ,,	3	0
Tenbury and Bewdley Railway (1864):				
Labourers	2	9 ,,	3	3
Wellington and Drayton Railway (1866):				
Labourers	2	9 ,,	3	3
Silverdale Railway (1868):				
Labourers	3	2 ,,	3	6

France.

Paris and Rouen Railway (1841):				
Labourers (English) . . .	3	6 to	3	9
,, (French) . . .	1	8 ,,	2	3

APPENDIX C.

Moldavia.

	Per diem.	
	s. d.	s. d.
At Jassy (1863):		
Common labourers	1 3 to	1 8
Masons		3 4
Carpenters		3 4
Blacksmiths		3 4
At Galatz (1863):		
Common labourers		1 3
Masons		3 6
Carpenters		3 6
Agricultural labourers		0 10[1]
(1870):		
Labourers		2 0
Carpenters		3 6
Masons		3 6

Saxony and Bohemia.

(1871):		
Labourers	2 0 to	2 6
Miners	2 0 ,,	2 6
Carpenters	2 6 ,,	3 0
Masons	2 6 ,,	3 0
Smiths (on Railways)	2 6 ,,	3 0
,, (in Ironworks)	2 0 ,,	2 3
,, (do., strikers)	1 3 ,,	1 5

Syria.

At Alexandretta:		
Common labourers		1 4
Masons	2 7 to	3 7
Carpenters	2 7 ,,	3 7
At Beyrout:		
Common labourers	2 0 to	2 6
Masons	3 0 ,,	5 0
Carpenters	3 0 ,,	5 0

[1] 6½d. in cash and 3½d. in food.

APPENDIX C.   199

	Per diem.	
	s. d.	s. d.
At Aleppo:		
Masons	2 3 to	2 9
„ labourers		1 10
„ boys		1 3
Carpenters	2 2 to	2 7

Persia.

(1870):		
Labourers		0 8
Masons	1 3 to	1 6
Carpenters	1 3 „	1 6
(Government Contracts):		
Labourers		0 6
Masons	0 10 „	1 0

India.

Delhi and Umritsir Railway (1864):		
Coolies	0 4½ to	0 6

Austria.

Lemberg and Czernowitz Railway (1864):		
Labourers	0 9 to	1 9
Carpenters		3 0
Masons		3 0
(1867):		
Labourers	1 6 to	1 8
Masons	2 0 „	2 6
Bricklayers	2 0 „	2 6
Carpenters		2 4
Smiths	2 0 „	2 6
(1871):		
Labourers	2 4 to	2 8
Masons	2 10 „	3 10
Bricklayers	2 10 „	3 10
Carpenters	2 6 „	2 10
Smiths	2 10 „	3 1

Belgium.

	Per diem.	
	s. d.	s. d.
Dutch Rhenish Railway (1852):		
Labourers	1 3 to	1 11
Carpenters		3 2
Masons		3 4

Italy.

Maremma Railway (1860):		
Labourers		0 10
Maremma Railway (1865):		
Labourers		1 8
Carpenters		2 6
Masons		2 6

Spain.

Bilbao and Miranda Railway (1858):		
Labourers		1 0
Masons		1 4
Bilbao and Miranda Railway (1863):		
Labourers		3 0
Masons		5 0

Canada.

Grand Trunk Railway (1852):		
Labourers	4 3 to	5 0
Masons	7 6 „	8 6
Carpenters	6 6 „	8 6

South America (Argentine Republic).

On the Coast (1864):		
Labourers	4 0 to	5 0
Carpenters	10 0 „	15 0
Masons	10 0 „	15 0
In the Interior:		
Labourers	2 0 to	3 0

APPENDIX C. 201

Australia.

	Per diem.	
	s. d.	s. d.
Queensland Railway,[1] (1863):		
Labourers	7 0 to	9 0
Masons	11 0 „	13 0
Bricklayers	11 0 „	12 0
Carpenters	10 0 „	12 0
Brickmakers	8 0 „	10 0

Hungary and Transylvania.

At Arad:		
Labourers		1 3
Masons	2 9 to	3 6
Carpenters	2 9 „	3 6
At Hermanstadt:		
Labourers		1 6
Masons[2]	3 3 to	4 6
Carpenters[2]	3 3 „	4 6
Smiths[2]	3 3 „	4 6

[1] 2,000 of the workmen employed on this railway had to be sent out from England, costing £5 per man for outfit and £12 per man to Government for passage out.

[2] These had to be brought from Bohemia.

INDEX.

A AND B, case, an, 162.
Accounts, Mr. Brassey's system of, 68.
Accuracy, Mr. Brassey's, 56.
Activity of Mr. Brassey, 62.
Admirer's opinion of Mr. Brassey, 82.
Advantages of English over foreign railway promoters, 30.
— derived from contract system, 22.
Agent, considering projects, 53; trust in, 5, 69, 125, 131; choice of, 5; list of, 84; relation between masters and, 163.
Aid from Government to railways, 105.
Alexander, Mr., 115.
America, Mr. Brassey goes to, 104; granaries in, *ib.*
'America,' the, off Cherbourg, 155.
American, granaries, 104; skill in machinery, 105.
Amount given away by Mr. Brassey, 81.
— expended on railways, 177.
"Anchor," ice, 111.
Anglo-Austrian Bank, 78.
Anxious and unanxious men, 6.
Appendices, frequently passed over, 83.
Appreciativeness of women, 167.
Arabic proverb, 80.
Argentine Railway, 132; account of the country, *ib.*; railway between Rosario and Cordova, 133; contractors for the line, 135; railway favourable to emigration, *ib.*; company's lands, 136; Rosario, *ib.*; Cordova, 137; soil and products, *ib.*; agricultural progress, *ib.*; sanitary conditions, 138; wages earned by immigrants, *ib.*; present political status, 139.
Armies of men, 38.
Armstrong, Sir W., and Victoria Docks, 120.
Australian works, 127-131.

BALLARD, MR., 63; meets Mr. Brassey, *ib.*; is employed on the G.N.R., 64; evidence of, 41, 65; selects men for Australia, 129.
Bank, Bilbao, 70.
Barentin Viaduct, fall of, 36; cost of, *ib.*; is rebuilt, 37.
Barrow Docks, loss on the, 77.
Bartlett, Mr., 74; superintends the Victor-Emmanuel Railway, 98; boring machine invented by, *ib.*; Mr Brassey aids, 99.
Basques, the, and paper money, 70.
Beattie, Mr., on the Crimean Railway, 120; dies, *ib.*
Beaumont, M., quotation from, on the Mont Cenis Tunnel, 99.
Belgians, the, as labourers, 47; taught "tipping" 48; their wages, 50.
Betts, Mr., superintends Grand

INDEX.

Trunk Railway, 104; a partner with Sir Morton Peto and Mr. Brassey, 101; organizes Crimean Railway, 119.
Bidder, Mr., 52; and the Victoria Docks, 120; Delhi Railway, 148.
Bilbao Bank, and paper money, 70.
— and Miranda Railway, 69; a Carlist sub-contractor's action on, 71; the difficulties in transporting money for the, 70.
Biography and History, 161.
Birkenhead, in 1818, 14; Mr. Lawton's ideas about, *ib.*; Mr. Harrison goes to, 17.
Birth of Mr. Brassey, 11.
Blame, reluctant, 6.
Bog, quaking, on G.N.R., how overcome, 64.
Bogie engines, 106.
Boring machine, 98.
Branborough Road, the, 84.
Brassey, Mrs., 18, 19, 173; advice to Mr. Brassey on railway matters, 18; marriage, 17; domestic labours of, 19.
Brassey, Mr., introduction of author to, 2; character of, 5; trustfulness, 2, 69; powers of calculation of, 52, 55, 56; of organization, 3; liberality, 7, 81, 164; equanimity, 5; perception, 6; delicacy in blaming, *ib.*; courtesy, 7; presence of mind, *ib*; hatred of contention, 8; anxiety to have work well done, *ib.*, 66; action in disputes, 8; ruling passion, 9, 164; birth, 11; parentage, *ib.*; family antecedents, *ib.*; goes to school, 12; is articled, *ib.*; a general favourite, 14; becomes Mr. Lawton's partner, *ib.*; Mr. Price's agent, *ib.*; introduced to Mr. George Stephenson, *ib.*; first tender, 15; meets Mr. Locke, 16; marriage, 17; moves house, 19; absence from home, *ib.*; fixes prices for sub-contractors, 25; choice of staff, 32; as a master, *ib.*; rapidity in executing work, 37, 43; power of dealing with schemes, 52; skill in estimating, *ib.*; mental arithmetic, *ib.*; projects laid before, *ib.*; confidence in agents, 53; would not enter Parliament, 63; dines with Napoleon III., 61; cheerfulness over financial loss, 74; credit attached to name of, 71; ignorance of his resources, 73; little love of money, 74; wealth, 80; amount of fortune, 82; an enthusiast's opinion of, *ib.*; meets Cavour, 95; a skilful financier, 97; goes to America, 104; and on to Canada, 108; reason for so doing, 109; love of engineering, 151; of nature, 152; a busy sightseer, *ib.*; taste for painting, 153; and sculpture, *ib.*; porcelain, 154; yachts, 155; Household brigade, *ib.*; hospitality, 156; oratory, *ib.*; politics, *ib.*; reading, 157; unworldliness, 164; mental refinement, 165; interview with Salamanca, 143; patience, 159; reproof, *ib.*; gentlemanlike conduct, 160; goes to Paris, &c., 169; correspondence, 55, 158; is taken ill, 171; failing health, *ib.*; death, 173; on Government railway control, 175.
Brassey, Mr. Albert, 173.
— Mr. Henry Arthur, M.P., 173.
— Mrs. Thomas, 170.
— Mr. Thomas, M.P., examiner

INDEX. 205

of evidence, 4; evidence about his father, 150.
Bravery, 7, 121, 167.
Bricks, 14, 44.
Bridges in Fen District, 65.
Buckhorn Western Tunnel, 55.
Buenos Ayres, the population of, 136.
Buerton, the Brasseys at, 11,
Buffalora Extension Railway, 99.
Bulkeley, 10.
Burgoyne, Field Marshal, 119.
Burnett, Dr., ix.; evidence, 158.
Business relations between Mr. Locke and Mr. Brassey, 16.
Butty-gangs, 27.

CALCULATION, Mr. Brassey's powers of, 52.
Canada, Mr. Brassey goes to, 109; Government of, lend money to the Grand Trunk Railway, *ib.*
— Works at Birkenhead, 105.
Canadians, Lower, on the Grand Trunk Railway, 108.
Capital, English and Italian railway, 93, 97; how Mr. Brassey employed his, 80.
"Captains of Industry," 3, 118.
Carl-Ludwig Railway, 78, 140.
Carlist, a, sub contractor, 71.
Case, Mr. Brassey states his, 3.
Cast-iron wheels, 105.
Caste, 147.
Cattle-plague, 179.
Cavour, Count, anxious to bring English capital into Piedmont, 93; communicates with Mr. Giles, *ib.*; meets Mr. Brassey, 95; meets Mr. Giles at Coire, *ib.*; reading Macaulay's History, 96; opinion of Mr. Brassey, 97.
Cenis, Mont, proposed tunnels through, 97; account of tunnel made, 98, 151.
Central control for railways, 177.

Chaffey's Mr., "Traveller," 113.
Character, Mr. Brassey's, 5.
Charleroi and Givet Railway, 35.
Cheerfulness of Mr. Brassey, 74.
Cherbourg, yachts at, 155.
Chicago, Mr. Brassey at, 152.
Child, the, and team of horses, 178.
Chivasso and Ivrea Railway, 99.
Choice of agents, 5.
Circumspice, 127.
Climate, of Italy, 46; of Argentine Republic, 138.
Coinage, 71.
Coire, Count Cavour and Mr. Giles's meeting at, 95; dinner at, 97.
Coleridge, on the beauty of feminine nature, 167.
Colonization and conquest, 133.
Comfort of foreign lines, 175.
Commercial training, 124.
Commissary-general, 38.
Common sense, 162.
Competition in railways, 177; in America, 105.
Concessions, cost of, 94; from Argentine Republic, 133; Moldavian, 141.
Confidence in subordinates, 163.
Conquest and colonization, 133.
Consideration of one's agents, 163.
Contractor, prejudice against the word, 3; drawbacks in the life of a, 19.
Contractors, need of, a mark of civilization, 21; as employers, 32.
Contracts, 21; first, 15; reason for, 22; sub-contracting, *ib.*, 24; advantages of, 22; government, 23; limits of, *ib.*; kind of work for, *ib.*; list of, 84.
Control, 175.
Conveyance of money in Spain, 70; in Austria, 78.
Co-operative system, 27.

INDEX.

Cordova, 137.
Correspondence, 55, 158.
Cost of Tunnel, 54: inspections, 26; of labour abroad and at home, 54; of Barentin Viaduct, 36; of Turin and Novara concession, 94; of Australian railways, 128; of living in Australia, 129; Indian lines, 148.
Council, of gangers, 8; of Indian chiefs, 112.
Courage, 121; "Two o'clock in the morning," 7.
Couza, Prince, 144.
Credit of Mr. Brassey's name, 71.
Crimean Railway, 118.
Cutting, men at work in, 38.

D'AZEGLIO, M., 97.
Danes, 123.
Danish contracts, 121; loss on, 76.
Darien, Isthmus of, 151.
Day, Mr., letter of, 185.
Death of Mr. Beattie, 120; of Mr. Brassey, 173.
Decimal coinage, 70.
Decisiveness, 123.
Delhi Railway, 148.
Dent, Mr., ix.
Details, attention to, 57.
Devotion of men, 172.
Difficulties in Spain, 69; in 1866, 75; in railway making, 18; in negotiating railways, 140; of English abroad, 33.
Disputes, Mr. Brassey's action in, 8.
Dixons, Messrs., aid Mr. Brassey, 15.
Docks, Barrow, 77; Victoria, 120.
Drunkenness, 122.
Dutton Viaduct, 15.

EAST London Railway, 121.
Eastern Bengal Railway, 147.
Economy, false, 118.
Education, 124.

Edwards, Mr., ix.
Elliot, Mr., M.P., 171.
Embarrassments in 1866, 75; on the Grand Trunk Railway, 109.
Emigrants, 130; railway making and, *ib.*; wages of, 138; cost of sending out to Australia, 129.
Emigration, 127, 133.
Engineers, difference between English and foreign, 45; of the various railways, 84.
Errors in railway promotion, 177.
Estimates, Mr. Locke keeping within, 30.
Eugénie, Empress, 61.
Evans, Mr., ix.
Evesham Railway, 76.
Evidence, how obtained for this work, 4.
— Mr. Ballard's, 41, 65.
— Mr. Thomas Brassey's, 150-160.
— Mr. Giles's, 93.
— Mr. Hawkshaw's, 21, 50, 191.
— Mr. Henfrey's, 99, 146.
— Mr. Hodges', 102.
— Mr. Jones's, 44.
— Mr. Mackay's, 42.
— Mr. Murton's, 30-37.
— Sir M. Peto's, 119.
— Mr. Rowan's, 102, 107, 123.
— Mr. Tapp's, 69, 181.
— Mr. Wilcox's, 128.
— Mr. Woolcott's, 136.
Excavators, 108.
Expenses, smallness of Mr. Brassey's personal, 82.
Exports of Argentine Republic, 137.

FAILURE, not unknown to Mr. Brassey, 75; from over-attention to details, 58; of Barentin Viaduct, 36.
Faults, 150.
Fell Railway, 151, 170.
Fête days, loss from, 72.

INDEX. 207

Financial difficulties, 75; on Grand Trunk Railway, in 1866, 109; on Danish railways, 121.
"Fly-tools," 46.
Fölsch, M., 181.
Foreign enterprises, 28.
Fortune, acquisition of a great, not pleasing, 80; amount of Mr. Brassey's, 82.
Fowler, Mr., ix.
Fox, Sir C., 151.
French railways compared with English, 175; and English engineers, 45.
Friends of Mr. Brassey, and author, 4.

GANGERS taken into council, 9.
General Credit Co. advance cash, 76.
Generosity, 81, 164.
Gentleman "of the old school," 3; living the life of a, 58.
Germans, 47.
Giles, Messrs. Brassey, Mill and, partners, 94.
Giles, Mr. Francis, 16.
Giles, Mr. Netlam, ix., 8, 60; evidence of, 93; meets Cavour at Coire, 95; on Moldavian lines, 141.
Good manners, 166.
Government aid to railways, 105; control, 175.
Granaries, American, 104.
Grand Junction Railway, 16.
Grand Trunk Railway, object of, 101; its rolling stock, 105; bogie engines adopted for, 106.
Graving dock, Thames, 120.
Great Britain, a fortunate country, 28.
"Great Eastern," ship, 151.
Great Eastern Railway shares, 77.
Great Northern Railway, 63; number of men on, 67.
Great Western branches, loss on, 76.

Ground ice, 111.

HANCOX, Mr., ix.
"Happy are, &c.," 80.
Harlings, Mr., Mr. Brassey's schoolmaster, 12.
Harrison, Mr., of Birkenhead, 18; predicts Mr. Brassey's successful career, *ib.*
Harrison, Mr. George, 115.
Harrison, Mr. Henry, 44; evidence, *ib.*, 121.
Harrison, Mr. Joseph, 148.
Hartland, Mr., and Victoria Docks, 120.
Hawkshaw, Mr., ix; quotation from, 21, 50; letter from, 191.
Heap, Mr., 115.
Henfrey, Mr., ix; his evidence, 24, 97, 98, 147.
Herz, Adolphe de, M., 140.
Hindoos, 146.
Hinks, the Hon. F., 101.
Hodges, Mr., ix; evidence, 102, 106.
Holme, Mr., letters, 186, 193.
Holyhead Road, 12.
Horses, not paying, 25; child and team of, 178.
Hospitality, Mr. Brassey's love of, 156.
Household brigade, 155.

ICE, on the St. Lawrence, 110; "anchor," 111; "ground," *ib.*
Idleness, 60.
India House, 118.
Indian lines, 146.
Indians, council of, 112.
Inspections, 6, 58; cost of, 26.
Interpreters, 33.
Interview, author's first, with Mr. Brassey, 2; between Cavour and Mr. Brassey, 95; Cavour and Mr. Giles, *ib.*; between Mr. Brassey and Salamanca, 143.
Introduction of author to Mr. Brassey, 2.

INDEX.

Inventions, 106.
Investigation of schemes, 52.
Iron Crown, 61; Order of the, given to Mr. Brassey, 79.
Ironwork, American skill in, 105.
Isidore, 182.
Italian railways, 93.
Italians employed in early ages, 22.

JACKSON, Sir William, partner with Messrs. Peto and Brassey, 98, 102.
Jones, Mr., ix; evidence, 44.

LABOUR, saving machinery, 104; scarcity of, 108; comparative value of, in England and abroad, 43; cost of, 54.
Labourers, lodgings for, 34; sent to Australia, 128; Piedmontese, 46; Neapolitan, ib.; German, 47; Belgian, ib.; Lower Canadian, 108; Danish, 123; Swedes, ib.; insufficiency of, in India, 148. See also Wages.
Lamb, Charles, 118.
Language, a navvy's, 33; number of languages spoken on a railway, 34.
Lavater, x.
Law courts, Mr. Brassey's name little known in, 8.
Lawton, Mr., Mr. Brassey articled to, 12; partners, 14.
Lea, Lord Herbert of, 3.
Legal expenses, 174.
Legion of Honour, 61.
Lemberg and Czernowitz Railway, 76, 144; finished before time, 77.
Letters, Mr. Brassey's, 55, 158; to Mr. Goodfellow, 185; Mr. Darke, 186; Mr. Day, 185; Mr. Holme, 186; Mr. Murton, 187; to Prince Sapieha, 142; from Mr.

Hawkshaw, 191; from Mr. Milroy, 187; from Mr. Holme, 193; from Mr. Wheelwright, 194.
Liberality, 5, 81.
"Lie there, Lord Treasurer," &c., 6.
Liverpool training, 125.
Locke, Mr., meets Mr. Brassey, 16; high appreciation of, 30; invited over to France, ib.
Lodgings for men, 34.
London and Southampton Railway, 16; directors of, applied to by the French, 30.
Longridge, Mr., ix.
Loss on the Bilbao Railway, 70; Mr. Brassey's cheerfulness over, 74; on Danish railways, 121.
Louth, Mr., ix.
Lower Canadians, 108.
Lucca, labourers from, 47.
Lukmanier Railway proposed, 95.

MACAULAY'S History, Cavour reading, 96.
McClean, Mr., 60, 141.
Machinery, saving labour, 104.
Mackay's, Mr., evidence, 42.
Mackenzie, Mr., joins partnership with Mr. Brassey, 31.
Mackintosh, Sir James, 2.
Mackintosh, Mr., 15.
Maison Carrée, the, 153.
Manby, Lieut.-Col., ix.
Manners, 166.
Maps:
— English Railways, 59.
— French and Italian Railways, 29.
— Grand Trunk Railway, 103.
— Argentine Railway, 134.
Market Drayton Railway, 77.
Marriage, Mr. Brassey's, 17.
Maus, Chevalier, 97.
Mavrojeny, M., 141.
Mechanics, two sent out for the

INDEX. 209

Canada Works, 105; as emigrants, 114.
Medail, M., 97.
Medical staff, 33.
Memory, 55.
Mental arithmetic, 55.
Mercantile training, 124.
Merits of others, 167.
Mid-Level Sewer, 120.
Mills, Mr., 93.
Milroy, Mr., ix; letter, 187.
Mind, management of the, 55.
Miners, 44.
Moldavian lines, 140.
Monetary difficulties in Spain, 70; Austria, 78; in 1866, 75.
Money, conveyance of, 78; paper, 71.
Mont Cenis Tunnel, M. Beaumont on the, 99.
Morticing machines, 106.
Murton, Mr., ix; evidence, 30-37; letter to, 187.

NANTWICH Railway, 77.
Napoleon III., Mr. Brassey dines with, 61; receives Legion of Honour from, *ib*.
Navvy, the English, abroad, 42; language, 33; conduct abroad, 42; dress, *ib*.; origin of name, 41; a phrase, 48; severity of their work, 41; their wives and families, 42; at work in a cutting, 38; recognition of, by Mr. Brassey, 172.
Neapolitans, 46; their system of labour, *ib*.; wages, *ib*.; food, 47.
Number of labourers employed on sub-contracts, 27; on Great Northern, 67; on Bilbao Railway, 71; on Victoria Bridge, 116; of men sent to Australia, 129.

OFENHEIM, Chevalier d', 78, 182.
Ogilvie, Mr., ix.
Old school, gentleman of the, 3.

Oratory, Mr. Brassey's love of, 156.
Organization, 38; powers of, 3, 6; want of, on railways, 175.

PALEOCAPA, M., 93.
Paper-money and the Basques, 71.
Parents, advice to, 158.
Paris fortifications, 27, 49.
Paris and Rouen Railway, 30; the directors apply to those of the Southampton Railway, *ib*.; number of languages spoken on, 34.
Parliament, Mr. Brassey refuses to enter, 60.
Parliamentary expenses, and Turin and Novara Railway, 94; committee, 174.
Partners, Mr. Brassey's, 84.
Patience, 159.
Pauperism in Australia, 130.
Paxton, Sir J., 147.
"Peas overgrowing the sticks," 7.
Peel, Sir Robert, on railways, 176.
Penkridge Viaduct, 15, 172.
Penson, Mr., 12.
Peto, Sir Morton, ix; evidence, 119; partnership with, 101.
Piedmontese, labourers, 46; the, public and Novara Railway, 95.
Pigeon English, 33.
Planing machines, 106.
Plate-laying, 44.
Poetry of engineering, 117.
Policy, foreign, in railway matters, and of the English Government, 45.
Politics of Mr. Brassey, 156.
Polytechnic schools, 123.
Population of Rosario, 136; of Cordova, 137.
Presence of mind, 71, 162.
Price, Mr., Mr. Brassey agent to, 14.
Produce of Rosario, 136; Cordova, 137.

P

INDEX.

Proverb, an Arabic, 80.
Purdon, Mr., 147.

QUAKING bog, a, 64.
Quantities, taking out, 53; of the Victoria Bridge, 116.
Quarries, Stourton, 14; Point St.-Claire, 112.
Queensland Railway, 76.
Quotation from "Life of Mackintosh," 2.
— "Géologie des Alpes," 99.
— Ralston, 179.
— Sydney Smith, 2.
— "Spanish Conquest," 244.

RAILWAY, Argentine (Central), 133.
— Australian, 126.
— Bilbao and Miranda, 69, 73.
— Carl-Ludwig, 140.
— Charleroi and Givet, 35.
— Cherbourg, 96.
— Chivasso and Ivrea, 99.
— Crimean, 118.
— Delhi, 24, 148.
— East London, 121.
— Eastern Bengal, 147.
— Evesham and Redditch, 76.
— Grand Trunk, 101.
— Great Eastern, 77.
— Great Northern, 63.
— Great Western branches, 76.
— Lemberg and Czernowitz, 76, 77, 140.
— London and Southampton, 16.
— Lukmanier, 95.
— Market Drayton, 77.
— Paris and Rouen, 30.
— Rosario, 133.
— Rouen and Havre, 35.
— Queensland, 76.
— Salisbury and Yeovil, 54.
— Sambre and Meuse, 35.
— Turin and Novara, 93.
— Turin and Susa, 97.
— Victor-Emmanuel, 98.
— Warsaw and Terespol, 76.
— Wellington and Drayton, 77.

Railway, Wolverhampton and Walsall, 172.
— General list of, Contracts, 84.
Railways, contrast between English and American, 104; early objections to, 18; respective position of England and France in regard to early, 28; organization in making, 38, 174; sums expended on, 177; English mode of making, 47; want of consideration in construction of, 174; and the South Sea Bubble, 175; neglect of public convenience on, ib.; Sir Robert Peel and, 176; influence over money matters, 70; as a means of communication, 104, 135; and Government control, 175; original skill in, not followed up, 174; in contrast with water transport, 104.
Rapidity of Mr. Brassey's execution of work, 37, 43.
Ray, Mr., ix.
Readers and writers have but one object, 1.
Reading, 157.
Red tape, Sydney Smith on, 2.
"Red 'un," a, 48.
Reikie, Mr., on Grand Trunk, 102.
Remuneration of Mr. Brassey's staff, 73.
Reports of agents, 53.
Retiring from business, 60.
Rhodes, Mr., ix.
Ricketts, Mr., ix, 83.
Road-making, 40.
Robinson, Rev. H., ix.
Rolling stock, 105.
Rosario, population of, 136; commerce of, ib.; and Cordova Railway, 133.
Ross, Mr. A., 102.
— the Honourable John, a promoter of the Grand Trunk, 102.

INDEX. 211

Rothschild, Baron Anselm de, 142.
Rouen, Mr. Brassey goes to, 36.
— and Havre Railway, 35.
Rowan, Mr., 102, 107; evidence, 122, 123; on British and Foreign labour, 107; on steam excavator, 108.
Ruling passion, 9, 164.
Runcorn Bridge, 77.

St.-Claire quarries, 112.
St. Lawrence, ice on the, 111; width of, *ib.*
Salamanca, Marquis, 143.
Salaries, 73.
Sambre and Meuse Railway, 35; payment of wages on, *ib.*
Sankey Viaduct, 15.
Sapieha, Prince, 141.
Schwarz, M., 181.
Scrivener, Dr., 138.
Season, shortness of, in Italy, 46; in Canada, 114.
"Set" of wagons, 41.
Shares, 143.
Shield, The Brassey, 67.
Skill in choice of men, 5.
Small's, Mr., adventures, 72.
Smith, Mr., and Mr. Ballard, 64.
Smith, Sydney, 2.
Sommeiller, M., 96; boring machine, 99.
Sons writing biography, 2.
Spain, Mr. Tapp's difficulties in, 69.
Speciality, each to one's own, 27.
Speeches, Mr. Henfrey's, 24; Mr. Stephenson's, 117.
Spence, Mr., the sculptor, 153.
"Spilt milk, crying over," 44.
Spoil banks, 49.
Staff, choice of, 32.
Staffholder sent for help on Bilbao line, 72.
State control over railways, 175.
Steam excavator, uses of, 108.
— traveller, 113.
Stephenson, Mr. George, Mr. Brassey meets, 14.

Stephenson, Mr. Robert, consulting engineer of Grand Trunk, 102; and the "poetry of engineering," 117.
Steppe cattle, 179.
Stileman, Mr., 141.
Stone for Victoria Bridge, 112.
Stores, railway, thrown open for Crimean Railway, 119.
Stourton quarries, 14.
Strapp, Mr., ix, 56.
Strikes, 113.
Sub-contractor, 27; Mr. Brassey's dealing with, 25; fixes prices for, *ib.*; a Carlist, 71.
Sub-contracts, 22, 24; amount of some, 27; sublet, *ib.*
"Suspect," a son's account of his father a little, 2.
Sweedish labourers, 123.
Swiss, 96.
System of French railway control, 175; sub-contracting, 22; of accounts, 68; of remuneration, 73.

Tait, Mr., an agent on the Grand Trunk Railway, 102.
Talabot, M., 142.
Tapp, Mr., ix; Secretary to Mr. Brassey, 68; monetary difficulties of, in Spain, 69; evidence, *ib.*; account of Mr. Brassey's journeys, 181.
Technical education, 123.
Temperature, 136.
Tender, first, 15; unsuccessful, *ib.*
Testimonial, 67.
Thames graving docks, 120.
Tipping, 48.
Tools, foreign, 43, 46.
Trades' Unions, 42; injurious effects of, Mr. Hodges on, 107.
Trent Valley contract, 44.
Trust in agents, 5, 69, 125, 131, 162.
Tunnel, cost of Buckhorn Western, 55.
Turin and Novara Railway, 93;

cost of concession, 94; a successful undertaking, 95.
Turin and Susa Railway, 97.
"Two o'clock in the morning" courage, 7.

UNDERSTATING a case, 3.

VIADUCT, Sankey, 15; Dutton, *ib*; Penkridge, 15, 172; Barentin, 36, 37.
Victor-Emmanuel Railway, 98.
Victoria Bridge, 110; Robert Stephenson reports on, *ib*.; construction of, 111; stone quarries, 112; floating dams used, *ib*.; strikes and sickness, 113; financial difficulties, *ib*.; Chaffey's traveller used, *ib*.; shortness of working season, 114; the iron work, 115; material used, 116; completed, *ib*.
— Docks, 120.

WAGES in America, 200; in Argentine Republic, *ib*.; in Australia, 201; in Austria, 199; in Belgium, 35, 200; in England, 197; of a navvy, *ib*.; in France, *ib*.; in Moldavia, 198; in Syria, *ib*.; in India, 199; in Italy, 47, 200; in Persia, 199; in Spain, 200; of Neapolitans, 47; of emigrants, 138; table of, 197.
Wagons ordered for Great Northern Railway, 25; a "set" of, 41.
Wagstaff, Mr., ix, 147; aids Mr. Brassey, 169.
War, evil effects of, 125.
Warsaw Railway, 76.
Wealth, Mr. Brassey's, 80; a cause of, 82; amount of, *ib*.
Wellington and Drayton Railway, 77.
Wheelwright, Mr., letter from, 194.
Whitworth's tools, 106.
Wilcox, Mr., ix; evidence, 128.
Wool, 137.
Woolcott, Mr., ix; evidence, 136.
Word, Mr. Brassey as good as his, 17.
Work, different modes of, 40; the outcome of the whole man, *ib*.
Workmen, the navvy, 41; platelayers, 44; miners, *ib*.; Neapolitans, 46; Germans, 47; Belgians, *ib*.
Wren, Sir Christopher, 127.
Writers, relation to readers, 1.
Wythes, Mr., 147.

www.ingramcontent.com/pod-product-compliance
Ingram Content Group UK Ltd.
Pitfield, Milton Keynes, MK11 3LW, UK
UKHW040657180125
453697UK00010B/235